Physical Methods in Chemistry and Nano Science.
Volume 5: Molecular and Solid State Structure

Physical Methods in Chemistry and Nano Science.
Volume 5: Molecular and Solid State Structure

Editor

Andrew R. Barron

Contributors

Aditya Agrawal, John J. Allen, Andrew R. Barron,
Tyler Boyd, Gabriela Escalera, Natalia Gonzalez Pech,
Wilhelm Kienast, Wayne Lin, Gladys A. López-Silva,
Pavan M. V. Raja, Alexis van Venrooy, Farrukh Vohidov,
Changsheng Xiang, Jing Han Yap

MiDAS Green Innovations
2020

Cover image © 2008 by goktugg

First Printing: 2020

ISBN 978-1-8380085-8-1   .

9 781838 008581

MiDAS Green Innovation, Ltd
Swansea, SA1 8RD, UK

www.midasgreeninnovation.com

Dedication

To all my collaborators who have expanded my research horizons, without you I would still be doing organometallic main group chemistry.

"I could sleep for invisible years, carrying a message of hope"
Marillion (1997)

Contents

Acknowledgements

I would like to thank all the contributors to this Volume, for their interest in creating a user-friendly text for their peers.

The myriad collaborators of whom I have had the pleasure of learning from over the years are to be thanked for bringing new characterization methods to my research - you know who you are.

Last, but not least, I would like to thank my wife, Merrie, for putting up with me during the Editing of this book while 'social distancing' during the COVID-19 pandemic.

Preface

This Series intended as a survey of research techniques used in modern chemistry, materials science, and nanoscience. The topics are grouped into volumes, not be method *per se*, but with regard to the type of information that can be obtained. Thus, the Volumes are ordered as follows:

- Elemental composition.
- Physical and thermal analysis.
- Chromatography
- Chemical speciation.
- Molecular and solid state structure.
- Surface morphology and structure at the nanoscale.
- Device performance.
- Applications of analytical methods

As a consequence of this organization methods can be found in different Volumes. For example, X-ray photoelectron spectroscopy is included under Elemental Composition (Volume 1) with regard to its use for determining the chemical composition, while it is included under Chemical Speciation (Volume 3) with regard to determining the identity of component chemical moieties.

The goal was to create simple to understand explanations of methods that allow the reader to gain the knowledge to correctly apply a technique or interpret data. As a consequence, the topics in this book have been developed in partnership with undergraduate and postgraduate students at Rice University over a 7-year period, and because of this there is some variation in depth and focus given to each topic. I make no apology for this diversity.

Chapter 1: X-ray Crystallography

Aditya Agrawal, John J. Allen, Tyler Boyd, Wayne Lin, Alexis van
Venrooy, Jing Han Yap and Andrew R. Barron

History of X-ray crystallography

The birth of X-ray crystallography is considered by many to be marked by the
formulation of the law of constant angles by Nicolaus Steno in 1669 (Figure
1.1). Although Steno is well known for his numerous principles regarding all
areas of life, this particular law dealing with geometric shapes and crystal
lattices is familiar ground to all chemists. It simply states that the angles between corresponding faces on crystals are the same for all specimens of the
same mineral. The significance of this for chemistry is that given this fact,
crystalline solids will be easily identifiable once a database has been estab-
lished. Much like solving a puzzle, crystal structures of heterogeneous
compounds could be solved very methodically by comparison of chemical
composition and their interactions.

**Figure 1.1: Danish pioneer in both anatomy and geology Nicolas Steno (1638 -
1686).**

Although Steno was given credit for the notion of crystallography, the man
that provided the tools necessary to bring crystallography into the scientific
arena was Wilhelm Roentgen (Figure 1.2), who in 1895 successfully pio-
neered a new form of photography, one that could allegedly penetrate through
paper, wood, and human flesh; due to a lack of knowledge of the specific

workings of this new discovery, the scientific community conveniently labeled the new particles X-rays. This event set off a chain reaction of experiments and studies, not all performed by physicists. Within one single month, medical doctors were using X-rays to pinpoint foreign objects such in the human body such as bullets and kidney stones (Figure 1.3).

Figure 1.2: German physicist Wilhelm Conrad Röentgen (1845 - 1923).

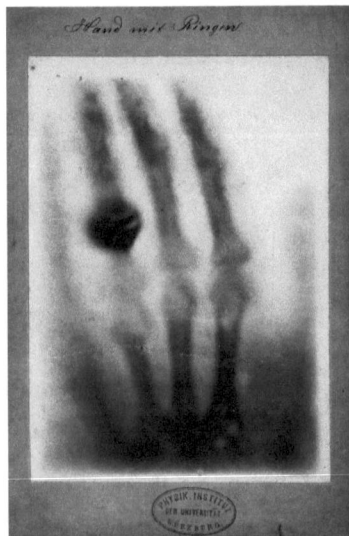

Figure 1.3: First public X-ray image ever produced. Pictured is the left hand of Anna Berthe Röentgen. The uncharacteristic bulge is her ring.

The credit for the actual discovery of X-ray diffraction goes to Max von Laue (Figure 1.4), to whom the Nobel Prize in physics in 1914 was awarded for the discovery of the diffraction of X-rays. Legend has it that the notion that eventually led to a Nobel prize was born in a garden in Munich, while von Laue was pondering the problem of passing waves of electromagnetic radiation through a specific crystalline arrangement of atoms. Because of the relatively large wavelength of visible light, von Laue was forced to turn his attention to another part of the electromagnetic spectrum, to where shorter wavelengths resided. Only a few decades earlier, Röentgen had publicly announced the discovery of X-rays, which supposedly had a wavelength shorter than that of visible light. Having this information, von Laue entrusted the task of performing the experimental work to two technicians, Walter Friedrich and Paul Knipping. The setup consisted of an X-ray source, which beamed radiation directly into a copper sulfate crystal housed in a lead box. Film was lined against the sides and back of the box, so as to capture the X-ray beam and its diffraction pattern. Development of the film showed a dark circle in the center of the film, surrounded by several extremely well-defined circles, which had formed as a result of the diffraction of the X-ray beam by the ordered geometric arrangement of copper sulfate. Max von Laue then proceeded to work out the mathematical formulas involved in the observed diffraction pattern, for which he was awarded the Nobel Prize in physics in 1914.

Figure 1.4: German physicist Max Theodor Felix von Laue (1879 - 1960) won the Nobel Prize for discovery of the diffraction of X-rays by crystals.

Principles of X-ray diffraction (XRD)

The simplest definition of diffraction is the irregularities caused when waves encounter an object. Diffraction is a phenomenon that exists commonly in everyday activities but is often disregarded and taken for granted. For example, when looking at the information side of a compact disc, a rainbow pattern will often appear when it catches light at a certain angle. This is caused by visible light striking the grooves of the disc, thus producing a rainbow effect (Figure 1.5), as interpreted by the observers' eyes. Another example is the formation of seemingly concentric rings around an astronomical object of significant luminosity when observed through clouds. The particles that make up the clouds diffract light from the astronomical object around its edges, causing the illusion of rings of light around the source. It is easy to forget that diffraction is a phenomenon that applies to all forms of waves, not just electromagnetic radiation. Due to the large variety of possible types of diffractions, many terms have been coined to differentiate between specific types. The most prevalent type of diffraction to X-ray crystallography is known as Bragg diffraction, which is defined as the scattering of waves from a crystalline structure.

Figure 1.5: The rainbow effects caused by visible light striking the grooves of a compact disc (CD).

Formulated by William Lawrence Bragg (Figure 1.6), the equation of Bragg's law relates wavelength to angle of incidence and lattice spacing,

$$n\lambda = 2d \sin(\theta)$$

where n is a numeric constant known as the order of the diffracted beam, λ is the wavelength of the beam, d denotes the distance between lattice planes, and θ represents the angle of the diffracted wave. The conditions given by this equation must be fulfilled if diffraction is to occur.

Figure 1.6: Australian-born British physicist Sir William Lawrence Bragg (1890 - 1971).

Because of the nature of diffraction, waves will experience either constructive (Figure 1.7) or destructive (Figure 1.8) interference with other waves. In the same way, when an X-ray beam is diffracted of a crystal, the different parts of the diffracted beam will have seemingly stronger energy, while other parts will have seemed to lost energy. This is dependent mostly on the wavelength of the incident beam, and the spacing between crystal lattices of the sample. Information about the lattice structure is obtained by varying beam wavelengths, incident angles, and crystal orientation. Much like solving a puzzle, a three-dimensional structure of the crystalline solid can be constructed by observing changes in data with variation of the aforementioned variables

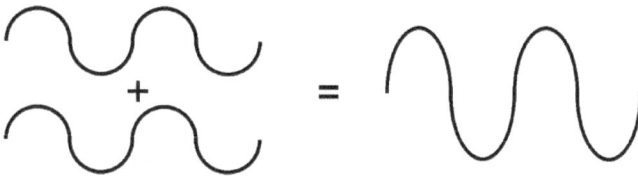

Figure 1.7: Schematic representation of constructive interference.

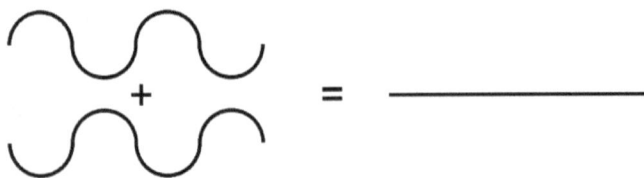

Figure 1.8: Schematic representation of destructive interference.

The X-ray diffractometer

At the heart of any XRD machine is the X-ray source. Modern day machines generally rely on copper metal as the element of choice for producing X-rays, although there are variations among different manufacturers. Because diffraction patterns are recorded over an extended period of time during sample analysis, it is very important that beam intensity remain constant throughout the entire analysis, or else faulty data will be procured. In light of this, even before an X-ray beam is generated, current must pass through a voltage regulator, which will guarantee a steady stream of voltage to the X-ray source.

Another crucial component to the analysis of crystalline via X-rays is the detector. When XRD was first developed, film was the most commonly used method for recognizing diffraction patterns. The most obvious disadvantage to using film is the fact that it has to be replaced every time a new specimen is introduced, making data collection a time-consuming process. Furthermore, film can only be used once, leading to an increase in cost of operating diffraction analysis.

Since the origins of XRD, detection methods have progressed to the point where modern XRD machines are equipped with semiconductor detectors, which produce pulses proportional to the energy absorbed. With these modern detectors, there are two general ways in which a diffraction pattern may be obtained. The first is called continuous scan, and it is exactly what the name implies. The detector is set in a circular motion around the sample, while a beam of X-ray is constantly shot into the sample. Pulses of energy are plotted with respect to diffraction angle, which ensure all diffracted X-rays are recorded. The second and more widely used method is known as step scan. Step scanning bears similarity to continuous scan, except it is highly computerized and much more efficient. Instead of moving the detector in a circle around the entire sample, step scanning involves collecting data at one fixed angle at a time, thus the name. Within these detection parameters, the types of detectors can themselves be varied. A more common type of detector, known as the charge-coupled device (CCD) detector (Figure 1.9), can be found in many XRD machines, due to its fast data collection capability. A CCD detector is comprised of numerous radiation sensitive grids, each linked to sensors that measure changes in electromagnetic radiation. Another commonly seen type of detector is a simple scintillation counter (Figure 1.10), which counts the intensity of X-rays that it encounters as it moves along a rotation axis. A comparable analogy to the differences between the two detectors mentioned would be that the CCD detector is able to see in two dimensions, while scintillation counters are only able to see in one dimension.

Figure 1.9: Single crystal X-ray diffractometer with a CCD detector. The large black circle at the left is the detector, and the X-ray beam comes out of the pointed horizontal nozzle. The beam stop can be seen across from this nozzle, as well as the gas cooling tube hanging vertically. The mounted crystal rests below the cooling gas supply, directly in the path of the beam. It extends from a glass fiber on a base (not shown) that attaches to the goniometer. The CCD detector can also be seen as the black tube on the right side of the photograph.

Aside from the above two components, there are many other variables involved in sample analysis by an XRD machine. As mentioned earlier, a steady incident beam is extremely important for good data collection. To further ensure this, there will often be what is known as a Söller slit or collimator found in many XRD machines. A Söller slit collimates the direction of the X-ray beam. In the collimated X-ray beam the rays are parallel, and therefore will spread minimally as they propagate (Figure 1.11). Without a collimator X-rays from all directions will be recorded; for example, a ray that has passed through the top of the specimen (see the red arrow in Figure 1.11a) but happens to be traveling in a downwards direction may be recorded at the bottom of the plate. The resultant image will be so blurred and indistinct as to be useless. Some machines have a Söller slit between the sample and the detector, which drastically reduces the amount of background noise, especially when analyzing iron samples with a copper X-ray source.

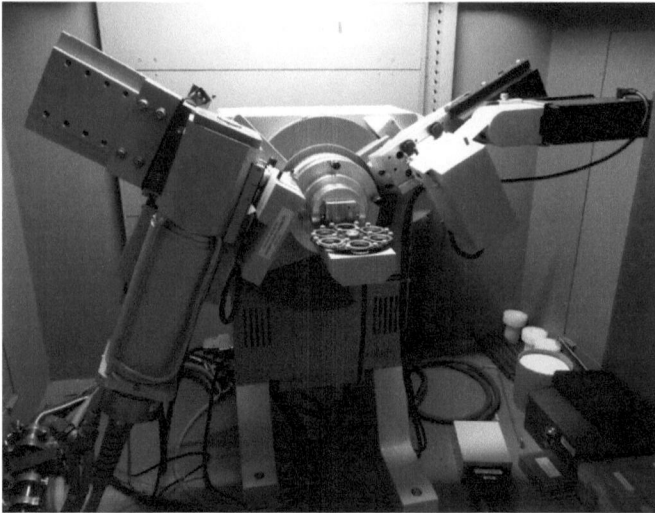

Figure 1.10: Image of a powder X-ray diffractometer. Two arms containing the X-ray source and detector are positioned around sample dishes, where the angle between each arm and the plane of the sample dishes is θ. The incident beam enters from the tube on the left, and the detector is housed in the black box on the right side of the machine. This particular XRD machine is capable of handling six samples at once and is fully automated from sample to sample.

This single crystal XRD machine (Figure 1.9) features a cooling gas line, which allows the user to bring down the temperature of a sample considerably below room temperature. Doing so allows for the opportunities for studies performed where the sample is kept in a state of extremely low energy, negating a lot of vibrational motion that might interfere with consistent data collection of diffraction patterns. Furthermore, information can be collected on the effects of temperature on a crystal structure. Also seen in Figure 1.9 is the hook-shaped object located between the beam emitter and detector. It serves the purpose of blocking X- rays that were not diffracted from being seen by the detector, drastically reducing the amount of unnecessary noise that would otherwise obscure data analysis.

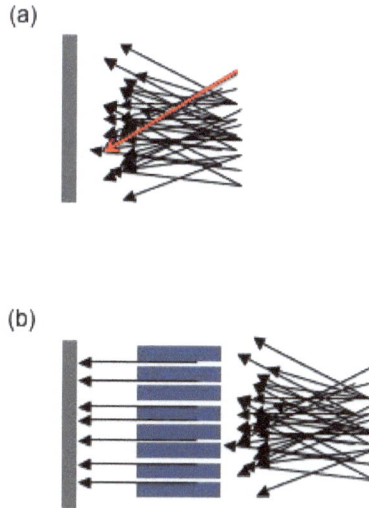

Figure 1.11: How a Söller collimator filters a stream of rays. (a) without a collimator and (b) with a collimator.

Evolution of powder XRD

Over time, XRD analysis has evolved from a very narrow and specific field to something that encompasses a much wider branch of the scientific arena. In its early stages, XRD was (with the exception of the simplest structures) confined to single crystal analysis, as detection methods had not advanced to a point where more complicated procedures were able to be performed. After many years of discovery and refining, however, technology has progressed to where crystalline properties (structure) of solids can be gleaned directly from a powder sample, thus offering information for samples that cannot be obtained as a single crystal. One area in which this is particularly useful is pharmaceuticals, since many of the compounds studied are not available in single crystal form, only in a powder.

Even though single crystal diffraction and powder diffraction essentially generate the same data, due to the powdered nature of the latter sample, diffraction lines will often overlap and interfere with data collection. This is apparently especially when the diffraction angle 2θ is high; patterns that emerge will be almost to the point of unidentifiable, because of disruption of individual diffraction patterns. For this particular reason, a new approach to interpreting powder diffraction data has been created.

There are two main methods for interpreting diffraction data:

- The first is known as the traditional method, which is very straight-forward, and bears resemblance to single crystal data analysis. This method involves a two-step process: 1) the intensities and diffraction patterns from the sample is collected, and 2) the data is analyzed to produce a crystalline structure. As mentioned before, however, data from a powdered sample is often obscured by multiple diffraction patterns, which decreases the chance that the generated structure is correct.
- The second method is called the direct-space approach. This method takes advantage of the fact that with current technology, diffraction data can be calculated for any molecule, whether or not it is the mol-ecule in question. Even before the actual diffraction data is collected, a large number of theoretical patterns of suspect molecules are gen-erated by computer and compared to experimental data. Based on correlation and how well the theoretical pattern fits the experimental data best, a guess is formulated to which compound is under question. This method has been taken a step further to mimic social interactions in a community. For example, first generation theoretical trial mole-cules, after comparison with the experimental data, are allowed to evolve within parameters set by researchers. Furthermore, if appro-priate, molecules are produced with other molecules, giving rise to a second generation of molecules, which fit the experimental data even better. Just like a natural environment, genetic mutations and natural selection are all introduced into the picture, ultimately giving rise a molecular structure that represents data collected from XRD analysis.

Another important aspect of being able to study compounds in powder form for the pharmaceutical researcher is the ability to identify structures in their natural state. A vast majority of drugs in this day and age are delivered through powdered form, either in the form of a pill or a capsule. Crystalliza-tion processes may often alter the chemical composition of the molecule (e.g., by the inclusion of solvent molecules), and thus marring the data if confined to single crystal analysis. Furthermore, when the sample is in powdered form, there are other variables that can be adjusted to see real-time effects on the molecule. Temperature, pressure, and humidity are all factors that can be changed in-situ to glean data on how a drug might respond to changes in those particular variables.

Powder X-Ray Diffraction

Powder X-Ray diffraction (XRD) was developed in 1916 by Debye (Figure 1.12) and Scherrer (Figure 1.13) as a technique that could be applied where traditional single-crystal diffraction cannot be performed. This includes cases where the sample cannot be prepared as a single crystal of sufficient size and quality. Powder samples are easier to prepare and is especially useful for pharmaceuticals research.

Figure 1.12: Dutch physicist and physical chemist Peter Joseph William Debye (1884 - 1966) recipient of the Nobel Prize in Chemistry.

Diffraction occurs when a wave meets a set of regularly spaced scattering objects, and its wavelength of the distance between the scattering objects are of the same order of magnitude. This makes X-rays suitable for crystallography, as its wavelength and crystal lattice parameters are both in the scale of angstroms (Å). Crystal diffraction can be described by Bragg diffraction,

$$\lambda = 2d \sin\theta$$

where λ is the wavelength of the incident monochromatic X-ray, d is the distance between parallel crystal planes, and θ the angle between the beam and the plane.

Figure 1.13: Swiss physicist Paul Scherrer (1890 - 1969).

For constructive interference to occur between two waves, the path length difference between the waves must be an integral multiple of their wavelength. This path length difference is represented by 2d sinθ Figure 1.14. Because sinθ cannot be greater than 1, the wavelength of the X-ray limits the number of diffraction peaks that can appear.

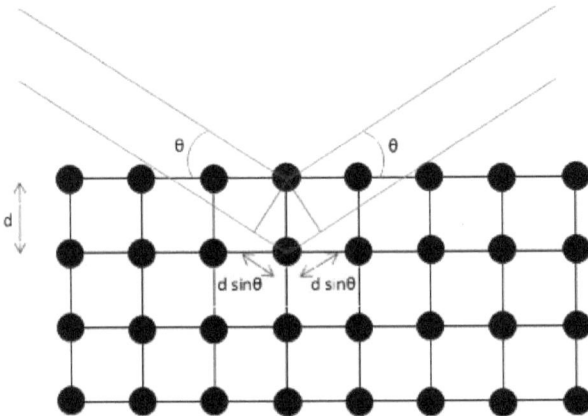

Figure 1.14: Bragg diffraction in a crystal. The angles at which diffraction occurs is a function of the distance between planes and the X-ray wavelength.

Production and detection of X-rays

Most diffractometers use Cu or Mo as an X-ray source, and specifically the K_α radiation of wavelengths of 1.54059 Å and 0.70932 Å, respectively. A stream of electrons is accelerated towards the metal target anode from a tungsten cathode, with a potential difference of about 30 - 50 kV. As this generates a lot of heat, the target anode must be cooled to prevent melting.

Detection of the diffracted beam can be done in many ways, and one common system is the gas proportional counter (GPC). The detector is filled with an inert gas such as argon, and electron-ion pairs are created when X-rays pass through it. An applied potential difference separates the pairs and generates secondary ionizations through an avalanche effect. The amplification of the signal is necessary as the intensity of the diffracted beam is very low compared to the incident beam. The current detected is then proportional to the intensity of the diffracted beam. A GPC has a very low noise background, which makes it widely used in labs.

Performing X-ray diffraction

Exposure to X-rays may have health consequences, follow safety procedures when using the diffractometer.

The particle size distribution should be even to ensure that the diffraction pattern is not dominated by a few large particles near the surface. This can be done by grinding the sample to reduce the average particle size to <10 μm. However, if particle sizes are too small, this can lead to broadening of peaks. This is due to both lattice damage and the reduction of the number of planes that cause destructive interference.

The diffraction pattern is actually made up of angles that did not suffer from destructive interference due to their special relationship described by Bragg Law (Figure 1.14). If destructive interference is reduced close to these special angles, the peak is broadened and becomes less distinct. Some crystals such as calcite ($CaCO_3$, Figure 1.15) have preferred orientations and will change their orientation when pressure is applied. This leads to differences in the diffraction pattern of 'loose' and pressed samples. Thus, it is important to avoid even touching 'loose' powders to prevent errors when collecting data.

The sample powder is loaded onto a sample dish for mounting in the diffractometer (Figure 1.10), where rotating arms containing the X-ray source and detector scan the sample at different incident angles. The sample dish is rotated horizontally during scanning to ensure that the powder is exposed evenly to the X-rays. A sample X-ray diffraction spectrum of germanium is shown in Figure 1.16, with peaks identified by the planes that caused that diffraction.

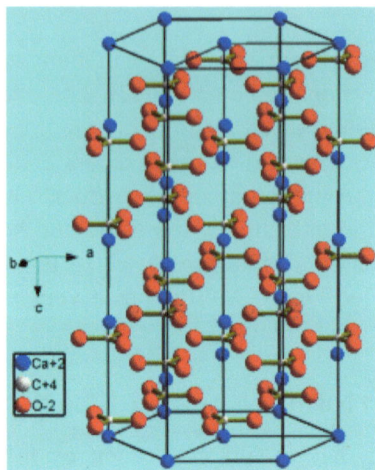

Figure 1.15: Calcite crystal structure. Under compression, the *c* axis orientates subparallel to the direction of pressure.

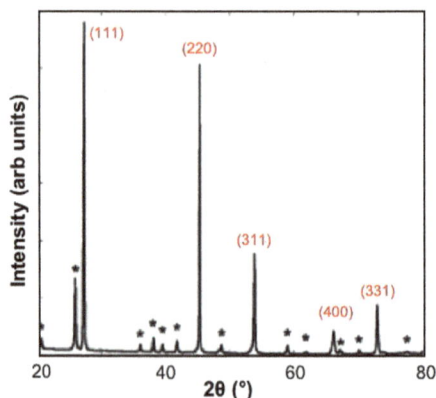

Figure 1.16: Powder XRD spectrum of germanium. Adapted from H. W. Chiu, C. N. Chervin, and S. M. Kauzlarich, Phase changes in Ge nanoparticles. *Chem. Mater.*, 2005, 17, 4858. Copyright: American Chemical Society (2013).

Germanium has a diamond cubic crystal lattice (Figure 1.17), named after the crystal structure of prototypical example. The crystal structure determines what crystal planes cause diffraction and the angles at which they occur. The angles are shown in 2θ as that is the angle measured between the two arms of the diffractometer, i.e., the angle between the incident and the diffracted beam (Figure 1.14).

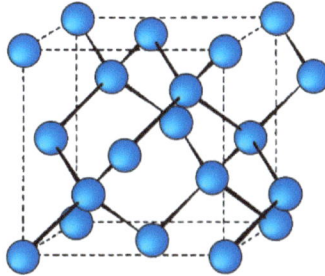

Figure 1.17: Model of diamond cubic crystal lattice.

Determining crystal structure for cubic lattices

There are three basic cubic crystal lattices, and they are the simple cubic (SC), body-centered cubic (BCC), and the face-centered cubic (FCC) Figure 1.18. These structures are simple enough to have their diffraction spectra analyzed without the aid of software.

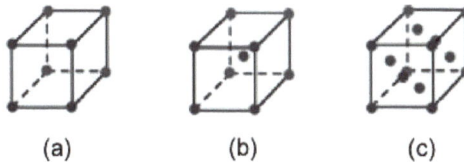

(a) (b) (c)

Figure 1.18: Models of cubic crystal structures.

Each of these structures has specific rules on which of their planes can produce diffraction, based on their Miller indices (hkl).

- SC lattices show diffraction for all values of (hkl), e.g., (100), (110), (111), etc.
- BCC lattices show diffraction when the sum of h+k+l is even, e.g., (110), (200), (211), etc.
- FCC lattices show diffraction when the values of (hkl) are either all even or all odd, e.g., (111), (200), (220), etc.
- Diamond cubic lattices like that of germanium are FCC structures with four additional atoms in the opposite corners of the tetrahedral

interstices. They show diffraction when the values of (hkl) are all odd or all even and the sum h+k+l is a multiple of 4, e.g., (111), (220), (311), etc.

The order in which these peaks appear depends on the sum of $h^2+k^2+l^2$. These are shown in Table 1.1.

(hkl)	$h^2+k^2+l^2$	BCC	FCC
100	1		
110	2	Y	
111	3		Y
200	4	Y	Y
210	5		
211	6	Y	
220	8	Y	Y
300, 221	9		
310	10	Y	
311	11		Y
222	12	Y	Y
320	13		
321	14	Y	
400	16	Y	Y
410, 322	17		
411, 330	18	Y	
331	19		Y
420	20	Y	Y
421	21		

Table 1.1: Diffraction planes and their corresponding $h^2+k^2+l^2$ values. The planes which result in diffraction for BCC and FCC structures are marked with a Y.

The value of d for each of these planes can be calculated using,

$$\frac{1}{d^2} = \frac{h^2 + k^2 + l^2}{a^2}$$

where a is the lattice parameter of the crystal. The lattice constant, or lattice parameter, refers to the constant distance between unit cells in a crystal lattice.

Example for NaCl

As the diamond cubic structure of Ge can be complicated, a simpler worked example for sample diffraction of NaCl with Cu-Kα radiation is as follows. Given the values of 2θ that result in diffraction, Table 1.2 can be constructed.

2θ (°)	θ (°)	Sinθ	Sin$^2\theta$
27.36	13.68	0.24	0.0559
31.69	15.85	0.27	0.0746
45.43	22.72	0.39	0.1491
53.85	26.92	0.45	0.2050
56.45	28.23	0.47	0.2237
66.20	33.10	0.55	0.2982
73.04	36.52	0.60	0.3541
75.26	37.63	0.61	0.3728

Table 1.2: Diffraction angles for NaCl.

Table 1.3 can then be constructed using various ratios of sin$^2\theta$ compared to the smallest value of sin$^2\theta$.

sin$^2\theta$	sin$^2\theta$/sin$^2\theta$	2×sin$^2\theta$/sin$^2\theta$	3×sin$^2\theta$/sin$^2\theta$	Corresponding (hkl)
0.0559	1.00	2.00	3.00	111
0.0746	1.33	2.67	4.00	200
0.1491	2.67	5.33	8.00	220
0.2050	3.67	7.34	11.00	311
0.2237	4.00	8.00	12.00	222
0.2982	5.33	10.67	16.00	400
0.3541	6.34	12.67	19.01	331
0.3728	6.67	13.34	20.01	420

Table 1.3: Ratio of diffraction angles for NaCl.

The values of these ratios can then be inspected to see if they corresponding to an expected series of hkl values. In this case, the last column gives a list of integers, which corresponds to the $h^2+k^2+l^2$ values of the FCC lattice diffraction. Hence, NaCl has a FCC structure, shown in Figure 1.19.

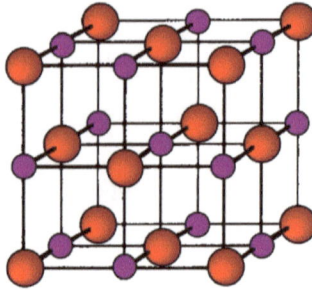

Figure 1.19: Model of NaCl FCC lattice.

The lattice parameter of NaCl can now be calculated from this data. The first peak occurs at $\theta = 13.68°$. Given that the wavelength of the Cu-K$_\alpha$ radiation is 1.54059 Å, Bragg's equation

$$\lambda = 2d \sin\theta$$

can be applied as follows:

$$1.54059 = 2d \sin(13.68)$$

$$d = 3.2571 \text{ Å}$$

Since the first peak corresponds to the (111) plane, the distance between two parallel (111) planes is 3.2571 Å. The lattice parameter can now be worked out using,

$$1/(3.2571)^2 = (1^2 + 1^2 + 1^2)/a^2$$

$$a = 5.6414 \text{ Å}$$

Determining composition

As seen above, each crystal will give a pattern of diffraction peaks based on its lattice type and parameter. These fingerprint patterns are compiled into databases such as the one by the Joint Committee on Powder Diffraction

Standard (JCPDS). Thus, the XRD spectra of samples can be matched with those stored in the database to determine its composition easily and rapidly.

Solid state reaction monitoring

Powder XRD is also able to perform analysis on solid state reactions such as the titanium dioxide (TiO$_2$) anatase to rutile transition. A diffractometer equipped with a sample chamber that can be heated can take diffractograms at different temperatures to see how the reaction progresses. Spectra of the change in diffraction peaks during this transition is shown in Figures 1.20 - 1.22.

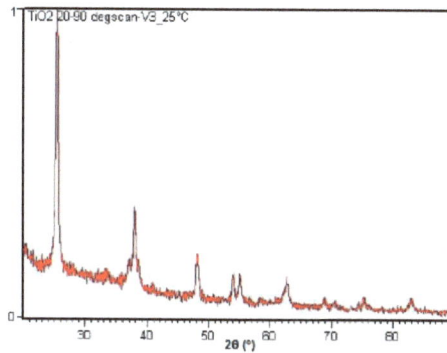

Figure 1.20: Powder XRD spectra of anatase TiO$_2$ at 25 °C. Courtesy of Jeremy Lee.

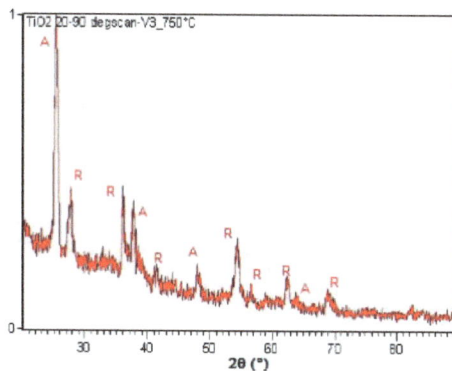

Figure 1.21: Powder XRD spectra of anatase (A) and rutile (R) TiO$_2$ at 750 °C, with labelled peaks for each phase. Courtesy of Jeremy Lee.

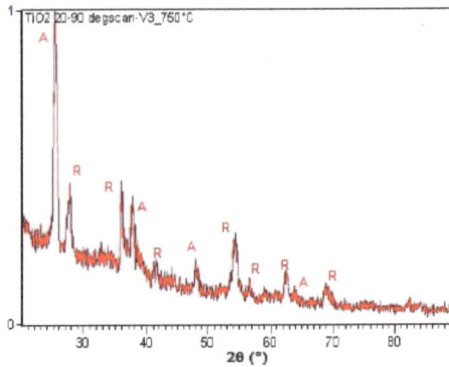

Figure 1.22: Powder XRD spectra of rutile TiO₂ at 1000 °C. Courtesy of Jeremy Lee.

XRD allows for quick composition determination of unknown samples and gives information on crystal structure. Powder XRD is a useful application of X-ray diffraction, due to the ease of sample preparation compared to single-crystal diffraction. Its application to solid state reaction monitoring can also provide information on phase stability and transformation.

Running a powder XRD of a complex mineral mixture

Evaluate and collect your sample.

The sample needs to be selected from the outcrop carefully. You need to make sure that the sample is representative and can answer the questions that you plan to ask. The sample must be disaggregated, which involves grinding the sample with a mortar and pestle to ensure a homogenous grain size. If the sample is not homogenous, the diffraction pattern of the mixture will be dominated by those of the large grains. Depending on the X-Ray source of your instrument, an ideal grain size for analysis is less than 10 μm. If a very fine grain size is desired, it is recommended to grind the sample while wet. This minimizes the opportunity for phase changes that can be induced by aggressive dry grinding.

The sample is loaded on the sample holder. It is important that there is enough sample present on the slide. This is because you want to have enough of each crystallite plane so that when the X-ray hits it at a particular angle, a peak will be produced (this increases the likelihood that Bragg's equation will be satisfied). Add the freshly ground powder to the slide and use a razor to scrape away any excess material that is not in the well. The sample holder is placed in the diffractometer. Often, a pressure plate is used to hold the sample in place. When collecting your data:

- Data should be collected from 2° - 100° (if using a Cu X-ray source).
- Step size should be 0.02°.
- When processing your data:
- Remove background.
- Smooth data.
- Identify peaks.

Compare your peaks to computer database for known mineral diffraction patterns. A useful database for mineral diffraction data is the open- access American Mineralogist Crystal Structure Database.

Experimental materials needed

- Sample.
- Agate mortar and pestle, a micronizing mill, or a mechanical grinder (the latter two are for harder samples) (Figure 1.23).
- Sample holder (If possible, a zero-background sample holder is preferred, especially when analyzing clays and other samples with weak diffraction patterns).
- Razor (used to cleanly load the sample onto the slide).
- Computer (for data analysis).

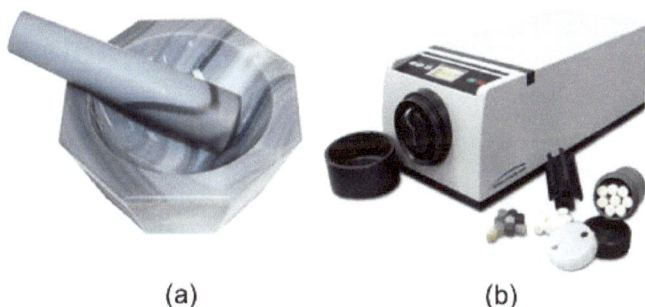

(a)   (b)

Figure 1.23: Agate mortar and pestle (a) and micronizing mill (b). Micronizing mill image adapted from McCrone group website (https://www.mccrone.com/mccrone-micronizing-mill).

Drawbacks and limitations

- XRD does have some drawbacks for the analysis of mineral samples.
- XRD does not provide quantitative data about the chemical composition of the mineral analyte.
- It does not provide textural data about the sample.
- Very minor components of a mineral mixture may not appear on the diffraction pattern.

Identification of clay minerals

Because different clay minerals are very similar in the x and y directions, the unit cell parameter along the z-direction is diagnostic (Figure 1.24). This poses a challenge for identifying clays, as the z unit cell parameter is the least pronounced. In addition, these weak diffraction signals are difficult to resolve from the background. Geologists must take special care when preparing clay samples for analysis. Because of this, further treatment and analysis of the clay should be done.

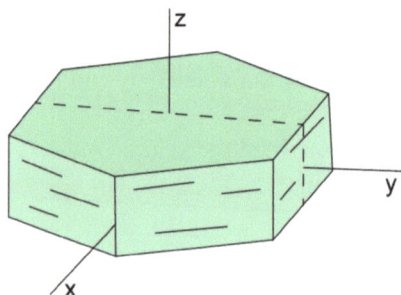

Figure 1.24: Schematic representation of the structure of a clay particle. Illustrates the diagnostic z axis used for X-ray crystallography. Image adapted from U.S. Geological Survey Open-File Report 01-041.
https://pubs.usgs.gov/of/2001/of01041/htmldocs/flow/flow.pdf.

Clay minerals belong to the class of minerals known as phyllosilicates, marked for their layered structures (Figure 1.25). Due to this, clay minerals can be altered between layers. These changes affect the crystal structure of the clay in the z-direction. The chemical and physical treatments and the corresponding diffraction pattern changes aid in the identification of the clay mineral. A detailed flowchart for this identification process is available online (USGS Open-File Report 01-041. https://pubs.usgs.gov/of/2001/of01041/htmldocs/flow/flow.pdf).

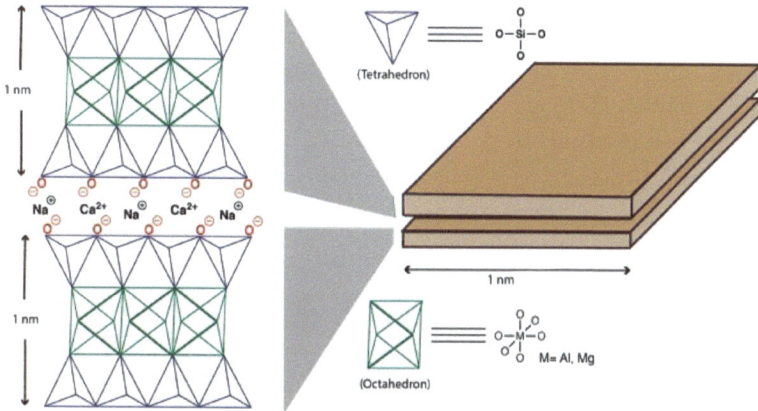

Figure 1.25: Structure of a clay mineral. Adapted from W. Zhu, C. Lu, F. Chang, and S. Kuo, Supramolecular ionic strength-modulating microstructures and properties of nacre-like biomimetic nanocomposites containing high loading clay. *RSC Adv.*, 2012, 2, 6295. Copyright: Royal Society of Chemistry (2012).

Determining the extent of aqueous alteration

CM-carbonaceous chondrites are primitive asteroids that have been incompletely serpentinized. They are of research interest because they contain information about nebular processes. They consist of Fe-Mg anhydrous silicates (olivine and pyroxene, see Figure 1.26) and hydrated phyllosilicates (serpentine family).

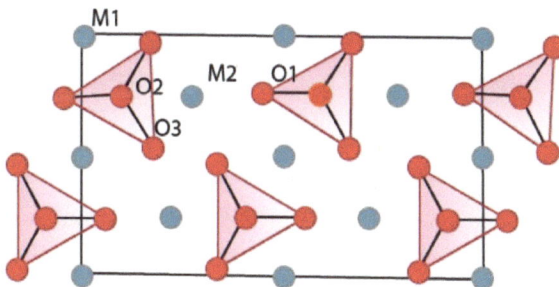

Figure 1.26: Example of olivine crystal structure, where SiO₂ tetrahedra shown in red and Mg/Al represented in blue. Oxygen sites and metal sites are indicated as M1, M2 (metal) and O1, O2, O3 (oxygen).

It has been shown that the conversion of Fe-chronsteditite to Mg-serpentine is proportional to the amount of aqueous alteration. Work has been done using position-sensitive detector XRD to determine the mineralogy of different CM chondrite samples (Figure 1.27). Knowing the extent of aqueous alteration will provide information about the source asteroid and may provide insight as to whether or not the meteorites came from the same parent body.

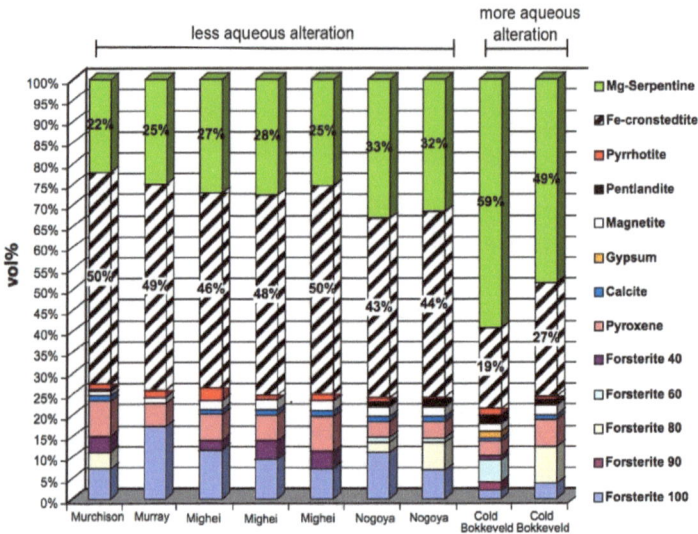

Figure 1.27: Modal mineralogy of CM chondrites determined using XRD. Chondrites with a higher percent Mg-Serpentine have experienced a greater amount of aqueous alteration. Reproduced from K. T. Howard, G. K. Benedix, P. A. Bland, and G. Cressey, Modal mineralogy of CM2 chondrites by X-ray diffraction (PSD-XRD). Part 1: Total phyllosilicate abundance and the degree of aqueous alteration. *Geochim. Cosmochim. Acta*, 2009, 73, 4576. Copyright: Elsevier (2009).

Identifying solid solutions and elemental substitutions

XRD can be used to identify solid solutions and elemental substitutions. In a recent example, the endmember phases of apatite-type alkaline earth element-yttrium-silicate oxybritholites were examined. Calcium, strontium, and barium are substituted into the crystal structure. Because the differences in the ionic radii of the cations are dramatic, the change affects the lattice structure. These changes are observed in the X-ray diffraction pattern (Figure 1.28 and Figure 1.29).

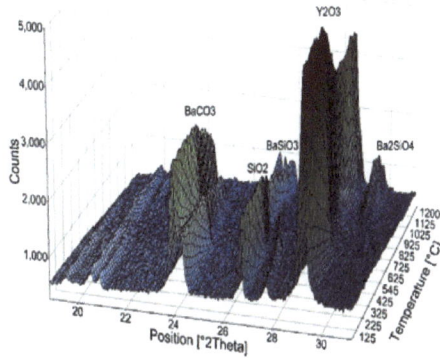

Figure 1.28: HT-XRD for the Ba-rich end phases of solid solution series of apatite-type alkaline earth element-yttrium-silicate oxybritholites. Reproduced from P. Ptacek, T. Opravil, F. Soukal, J. Tkacz, J. Masilko, and E. Bartonickova, The field of solid solutions in ternary system of synthetic apatite-type alkaline earth element-yttrium-silicate oxybritholite phases of the composition: $AEE_{\delta}Y_{10-\delta}[SiO_4]_6O_{3-0.5\delta}$, where AEE=Ca, Sr and Ba. *Ceram. Int.*, 2016, 42, 6154. Copyright: Elsevier (2016).

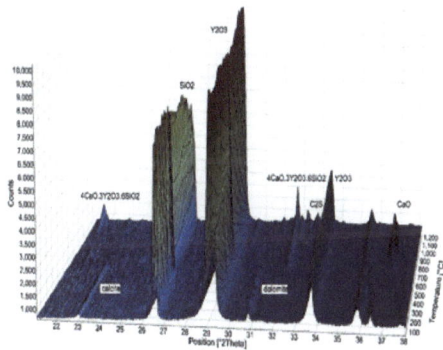

Figure 1.29: HT-XRD for the Ca-rich end phases of solid solution series of apatite-type alkaline earth element-yttrium-silicate oxybritholites. Reproduced from P. Ptacek, T. Opravil, F. Soukal, J. Tkacz, J. Masilko, and E. Bartonickova, The field of solid solutions in ternary system of synthetic apatite-type alkaline earth element-yttrium-silicate oxybritholite phases of the composition: $AEE_{\delta}Y_{10-\delta}[SiO_4]_6O_{3-0.5\delta}$, where AEE=Ca, Sr and Ba. *Ceram. Int.*, 2016, 42, 6154. Copyright: Elsevier (2016).

Using XRD of minerals for the determination of paleo-environmental conditions

In a 2008 study, XRD is used to learn the exact clay mineralogy of layers of lake sediments in the Swiss Alps. Paleo-climatic conditions have a direct effect on weathering patterns that, in turn, transport different clay minerals. The sediment mineralogy at the catchment will provide information about the weathering event that was required to transport a particular clay.

Single-crystal X-ray crystallography

Described simply, single-crystal X-ray diffraction (XRD) is a technique in which a crystal of a sample under study is bombarded with an X-ray beam from many different angles, and the resulting diffraction patterns are measured and recorded. By aggregating the diffraction patterns and converting them via Fourier transform to an electron density map, a unit cell can be constructed which indicates the average atomic positions, bond lengths, and relative orientations of the molecules within the crystal.

Fundamental principles

As an analogy to describe the underlying principles of diffraction, imagine shining a laser onto a wall through a fine sieve. Instead of observing a single dot of light on the wall, a diffraction pattern will be observed, consisting of regularly arranged spots of light, each with a definite position and intensity. The spacing of these spots is inversely related to the grating in the sieve the finer the sieve, the farther apart the spots are, and the coarser the sieve, the closer together the spots are. Individual objects can also diffract radiation if it is of the appropriate wavelength, but a diffraction pattern is usually not seen because its intensity is too weak. The difference with a sieve is that it consists of a grid made of regularly spaced, repeating wires. This periodicity greatly magnifies the diffraction effect because of constructive interference. As the light rays combine amplitudes, the resulting intensity of light seen on the wall is much greater because intensity is proportional to the square of the light's amplitude.

To apply this analogy to single-crystal XRD, we must simply scale it down. Now the sieve is replaced by a crystal and the laser (visible light) is replaced by an X-ray beam. Although the crystal appears solid and not grid-like, the molecules or atoms contained within the crystal are arranged periodically, thus producing the same intensity-magnifying effect as with the sieve.

Because X-rays have wavelengths that are on the same scale as the distance between atoms, they can be diffracted by their interactions with the crystal lattice.

These interactions are dictated by Bragg's law, which says that constructive interference occurs only when is satisfied (see Figure 1.14). A complication arises, however, because crystals are periodic in all three dimensions, while the sieve repeats in only two dimensions. As a result, crystals have many different diffraction planes extending in certain orientations based on the crystal's symmetry. For this reason, it is necessary to observe diffraction patterns from many different angles and orientations of the crystal to obtain a complete picture of the reciprocal lattice.

The reciprocal lattice of a lattice (Bravais lattice) is the lattice in which the Fourier transform of the spatial wavefunction of the original lattice (or direct lattice) is represented. The reciprocal lattice of a reciprocal lattice is the original lattice. The reciprocal lattice is related to the crystal lattice just as the sieve is related to the diffraction pattern: they are inverses of each other. Each point in real space has a corresponding point in reciprocal space and they are related by 1/d; that is, any vector in real space multiplied by its corresponding vector in reciprocal space gives a product of unity. The angles between corresponding pairs of vectors remains unchanged.

Real space is the domain of the physical crystal, i.e., it includes the crystal lattice formed by the physical atoms within the crystal. Reciprocal space is, simply put, the Fourier transform of real space; practically, we see that diffraction patterns resulting from different orientations of the sample crystal in the X-ray beam are actually two-dimensional projections of the reciprocal lattice. Thus, by collecting diffraction patterns from all orientations of the crystal, it is possible to construct a three-dimensional version of the reciprocal lattice and then perform a Fourier transform to model the real crystal lattice.

Technique

Single-crystal versus powder diffraction

Two common types of X-ray diffraction are powder XRD and single-crystal XRD, both of which have particular benefits and limitations. While powder XRD has a much simpler sample preparation, it can be difficult to obtain structural data from a powder because the sample molecules are randomly oriented in space; without the periodicity of a crystal lattice, the signal-to-noise ratio is greatly decreased, and it becomes difficult to separate reflections coming from the different orientations of the molecule. The advantage of powder XRD is that it can be used to quickly and accurately identify a known substance, or to verify that two unknown samples are the same material.

Single-crystal XRD is much more time and data intensive, but in many fields, it is essential for structural determination of small molecules and macromolecules in the solid state. Because of the periodicity inherent in crystals, small signals from individual reflections are magnified via constructive interference. This can be used to determine exact spatial positions of atoms in molecules and can yield bond distances and conformational information. The difficulty of single-crystal XRD is that single crystals may be hard to obtain, and the instrument itself may be cost-prohibitive.

An example of typical diffraction patterns for single-crystal and powder XRD follows (Figure 1.30 and Figure 1.31, respectively). The dots in the first image correspond to Bragg reflections and together form a single view of the molecule's reciprocal space. In powder XRD, random orientation of the crystals means reflections from all of them are seen at once, producing the observed diffraction rings that correspond to particular vectors in the material's reciprocal lattice.

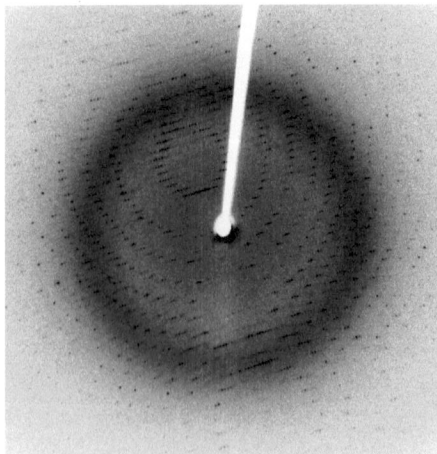

Figure 1.30: Single-crystal diffraction pattern of an enzyme. The white rod protruding from the top is the beam stop. Copyright: Je Dahl (2006).

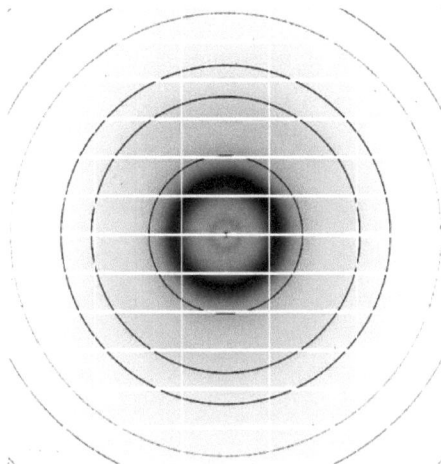

Figure 1.31: Powder X-ray diffraction spectrum of silicon. Taken by XanaG.

Method

In a single-crystal X-ray diffraction experiment, the reciprocal space of a crystal is constructed by measuring the angles and intensities of reflections in observed diffraction patterns. These data are then used to create an electron density map of the molecule which can be refined to determine the average bond lengths and positions of atoms in the crystal.

Instrumentation

The basic setup for single-crystal XRD consist of an X-ray source, a collimator to focus the beam, a goniometer to hold and rotate the crystal, and a detector to measure and record the reflections. Instruments typically contain a beamstop to halt the primary X-ray beam from hitting the detector, and a camera to help with positioning the crystal. Many also contain an outlet connected to a cold gas supply (such as liquid nitrogen) in order to cool the sample crystal and reduce its vibrational motion as data is being collected. A typical instrument is shown in Figure 1.9 and Figure 1.32.

Obtaining single crystals

Despite advances in instrumentation and computer programs that make data collection and solving crystal structures significantly faster and easier, it can still be a challenge to obtain crystals suitable for analysis. Ideal crystals are single, not twinned, clear, and of sufficient size to be mounted within the X-ray beam (usually 0.1-0.3 mm in each direction). They also have clean faces and smooth edges. Following are images of some ideal crystals (Figure 1.33 and Figure 1.34), as well as an example of twinned crystals (Figure 1.35). Crystal twinning occurs when two or more crystals share lattice points in a symmetrical manner. This usually results in complex diffraction patterns which are difficult to analyze and construct a reciprocal lattice.

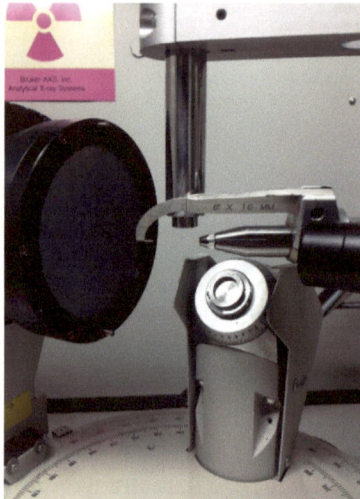

Figure 1.32: Close-up view of a single-crystal X-ray diffraction instrument. The large black circle at the left is the detector, and the X-ray beam comes out of the pointed horizontal nozzle. The beam stop can be seen across from this nozzle, as well as the gas cooling tube hanging vertically. The mounted crystal rests below the cooling gas supply, directly in the path of the beam. It extends from a glass fiber on a base (not shown) that attaches to the goniometer. The camera can also be seen as the black tube on the right side of the photograph.

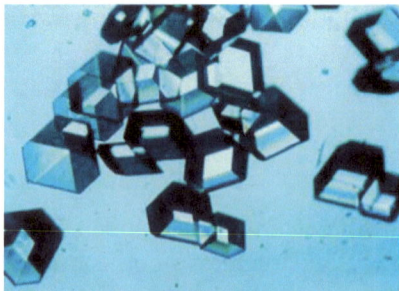

Figure 1.33: Single crystals of insulin grown in space; taken by NASA. Released under PD license.

Figure 1.34: An octahedral-shaped single crystal of synthetic chrome alum. Copyright: Ra'ike (2008).

Figure 1.35: Twinned quartz crystal. Image used under fair use license from the Geology, Gems, and Minerals Subject Guide of the Smithsonian National Museum of Natural History.

Crystal formation can be affected by temperature, pressure, solvent choice, saturation, nucleation, and substrate. Slow crystal growth tends to be best, as rapid growth creates more imperfections in the crystal lattice and may even lead to a precipitate or gel. Similarly, too many nucleation sites (points at which crystal growth begins) can lead to many small crystals instead of a few, well-defined ones.

There are a number of basic methods for growing crystals suitable for single-crystal XRD:

- The most basic method is to slowly evaporate a saturated solution until it becomes supersaturated and then forms crystals. This often works well for growing small-molecule crystals; macroscopic molecules (such as proteins) tend to be more di cult.
- A solution of the compound to be crystallized is dissolved in one solvent, then a 'non-solvent' which is miscible with the first but in which the compound itself is insoluble, is carefully layered on top of the solution. As the non-solvent mixes with the solvent by diffusion, the solute molecules are forced out of solution and may form crystals.
- A crystal solution is placed in a small open container which is then set in a larger closed container holding a volatile non-solvent. As the volatile non-solvent mixes slowly with the solution by vapor diffusion, the solute is again forced to come out of solution, often leading to crystal growth.
- All three of the previous techniques can be combined with seeding, where a crystal of the desired type to be grown is placed in the saturated solution and acts as a nucleation site and starting place for the crystal growth to begin. In some cases, this can even cause crystals to grow in a form that they would not normally assume, as the seed can act as a template that might not otherwise be followed.
- The hanging drop technique is typically used for growing protein crystals. In this technique, a drop of concentrated protein solution is suspended (usually by dotting it on a silicon-coated microscope slide) over a larger volume of the solution. The whole system is then sealed, and slow evaporation of the suspended drop causes it to become supersaturated and form crystals. (A variation of this is to have the drop of protein solution resting on a platform inside the closed system instead of being suspended from the top of the container.)

These are only the most common ways that crystals are grown. Particularly for macromolecules, it may be necessary to test hundreds of crystallization conditions before a suitable crystal is obtained. There now exist automated techniques utilizing robots to grow crystals, both for obtaining large numbers of single crystals and for performing specialized techniques (such as drawing a crystal out of solution) that would otherwise be too time-consuming to be of practical use.

Wide angle X-ray diffraction studies of liquid crystals

Liquid crystals

Some organic molecules display a series of intermediate transition states between solid and isotropic liquid states (Figure 1.36) as their temperature is raised. These intermediate phases have properties in between the crystalline solid and the corresponding isotropic liquid state, and hence they are called liquid crystalline phases. Other name is mesomorphic phases where mesomorphic means of intermediate form. According to the physicist de Gennes (Figure 1.37), liquid crystal is *an intermediate phase, which has liquid like order in at least one direction and possesses a degree of anisotropy*. It should be noted that all liquid crystalline phases are formed by anisotropic molecules (either elongated or disk-like) but not all the anisotropic molecules form liquid crystalline phases.

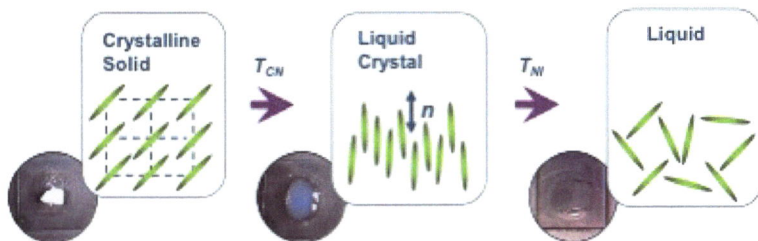

Figure 1.36: Schematic phase behavior for a molecule that displays a liquid crystal (LC) phase. T_{CN} and T_{NI} represents phase transition temperatures from crystalline solid to LC phase and LC to isotropic liquid phase, respectively.

Figure 1.37: French physicist and the Nobel Prize laureate Pierre-Gilles de Gennes (1932 - 2007).

Anisotropic objects can possess different types of ordering giving rise to different types of liquid crystalline phases (Figure 1.38).

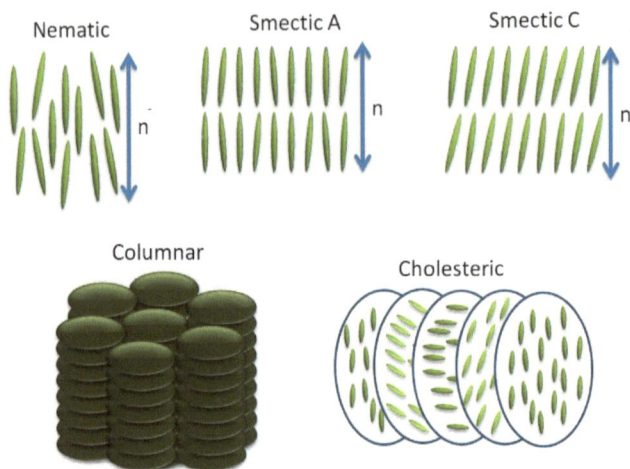

Figure 1.38: Schematic illustration of the different types of liquid crystal phases.

Nematic phases

The word *nematic* comes from the Greek for *thread* and refers to the thread-like defects commonly observed in the polarizing optical microscopy of these molecules. They have no positional order only orientational order, i.e., the molecules all pint in the same direction. The direction of molecules denoted by the symbol n commonly referred as the *director* (Figure 1.38). The director *n* is bidirectional that means the states *n* and -*n* are indistinguishable.

Smectic phases

All the smectic phases are layered structures that usually occur at slightly lower temperatures than nematic phases. There are many variations of smectic phases, and some of the distinct ones are as follows:
- Each layer in smectic A is like a two-dimensional liquid, and the long axis of the molecules is typically orthogonal to the layers (Figure 1.38). Just like nematics, the state *n* and -*n* are equivalent. They are made up of achiral and non-polar molecules.
- As with smectic A, the smectic C phase is layered, but the long axis of the molecules is not along the layers normal. Instead it makes an

angle (θ, Figure 1.38). The tilt angle is an order parameter of this phaseandcanvaryfrom0° to 45-50°.

- Smectic C* phases are smectic phases formed by chiral molecules. This added constraint of chirality causes a slight distortion of the Smectic C structure. Now the tilt direction precesses around the layer normal and forms a helical configuration.

Cholesterics phases

Sometimes cholesteric phases (Figure 1.38) are also referred to as chiral nematic phases because they are similar to nematic phases in many regards. Many derivatives of cholesterol exhibit this type of phase. They are generally formed by chiral molecules or by doping the nematic host matrix with chiral molecules. Adding chirality causes helical distortion in the system, which makes the director, n, rotate continuously in space in the shape of a helix with specific pitch. The magnitude of pitch in a cholesteric phase is a strong function of temperature.

Columnar phases

In columnar phases liquid crystals molecules are shaped like disks as opposed to rod-like in nematic and smectics liquid crystal phases. These disk-shaped molecules stack themselves in columns and form a 2D crystalline array structures (Figure 1.38). This type of two-dimensional ordering leads to new mesophases.

2D X-ray diffraction

X-ray diffraction (XRD) is one of the fundamental experimental techniques used to analyze the atomic arrangement of materials. The basic principle behind X-ray diffraction is Bragg's Law (Figure 1.14). According to this law, X-rays that are reflected from the adjacent crystal planes will undergo constructive interference only when the path difference between them is an integer multiple of the X-ray's wavelength.

Now the atomic arrangement of molecules can go from being extremely ordered (single crystals) to random (liquids). Correspondingly, the scattered X-rays form specific diffraction patterns particular to that sample. Figure 1.39 shows the difference between X-rays scattered from a single crystal and a polycrystalline (powder) sample. In case of a single crystal the diffracted rays point to discrete directions (Figure 1.39a), while for polycrystalline sample diffracted rays form a series of diffraction cones (Figure 1.39b). The X-ray patterns obtained are shown in Figures 1.30 and 1.31.

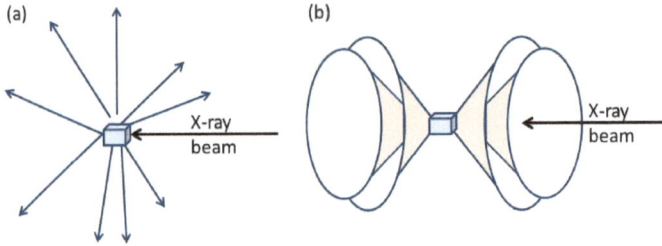

Figure 1.39: Schematic of the diffraction pattern from (a) single crystal and (b) polycrystalline sample

A two-dimensional (2D) XRD system is a diffraction system with the capability of simultaneously collecting and analyzing the X-ray diffraction pattern in two dimensions. A typical 2D XRD setup consists of five major components (Figure 1.40):

- X-ray source
- X-ray optics.
- Goniometer.
- Sample alignment and monitoring device.
- 2D area detector.

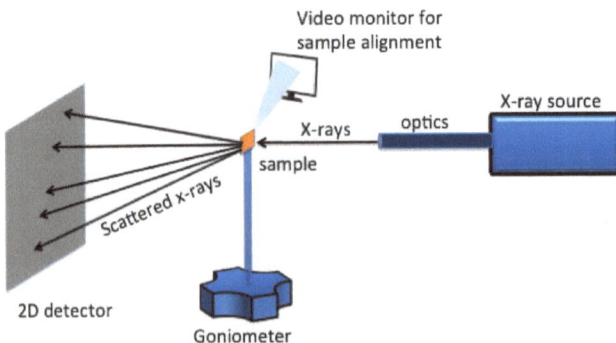

Figure 1.40: Schematic illustration of basic 2D WAXD setup. Adapted from B. B. He, U. Preckwinkel, and K. L. Smith, Fundamentals of two-dimensional X-ray diffraction (XRD 2). *Adv. X-ray Anal.***, 2000, 43, 273. Copyright: EVISA (2000).**

For laboratory scale X-ray generators, X-rays are emitted by bombarding metal targets with high velocity electrons accelerated by strong electric field in the range 20 - 60 kV. Different metal targets that can be used are chromium (Cr), cobalt (Co), copper (Cu), molybdenum (Mo) and iron (Fe). The most commonly used ones are Cu and Mo. Synchrotrons are even higher energy radiation sources. They can be tuned to generate a specific wavelength and they have much brighter luminosity for better resolution. Available synchrotron facilities in US are:

- Stanford Synchrotron Radiation Lightsource (SSRL), Stanford, CA.
- Synchrotron Radiation Center (SRC), University of Wisconsin-Madison, Madison, WI.
- Advanced Light Source (ALS), Lawrence Berkeley National, Berkeley, CA.
- National Synchrotron Light Source (NSLS), Brookhaven National Laboratory, Upton, NY.
- Advanced Photon Source (APS), Argonne National Laboratory, Argonne, IL.
- Center for Advanced Microstructures & Devices, Louisiana State University, Baton Rouge, LA.
- Cornell High Energy Synchrotron Source (CHESS), Cornell, Ithaca, NY.

The X-ray optics are comprised of the X-ray tube, monochromator, pinhole collimator and beam stop. A monochromator is used to get rid of unwanted X-ray radiation from the X-ray tube. A diffraction from a single crystal can be used to select a specific wavelength of radiation. Typical materials used are pyrolytic graphite and silicon. Monochromatic X-ray beams have three components: parallel, convergent and divergent X-rays. The function of a pinhole collimator is to filter the incident X-ray beam and allow passage of parallel X-rays. A 2D X-ray detector can either be a film or a digital detector, and its function is to measure the intensity of X-rays diffracted from a sample as a function of position, time, and energy.

Advantages of 2D XRD as compared to 1D XRD.

2D diffraction data has much more information in comparison diffraction pattern, which is acquired using a 1D detector. Figure 1.41 shows the diffraction pattern from a polycrystalline sample. For illustration purposes only, two diffraction cones are shown in the schematic. In the case of 1D X-ray diffraction, measurement area is confined within a plane labeled as diffractometer plane.

The 1D detector is mounted along the detection circle and variation of diffraction pattern in the z direction are not considered. The diffraction pattern collected is an average over a range defined by a beam size in the z direction. The diffraction pattern measured is a plot of X-ray intensity at different 2θ angles. For 2D X-ray diffraction, the measurement area is not limited to the diffractometer plane. Instead, a large portion of the diffraction rings are measured simultaneously depending on the detector size and position from the sample.

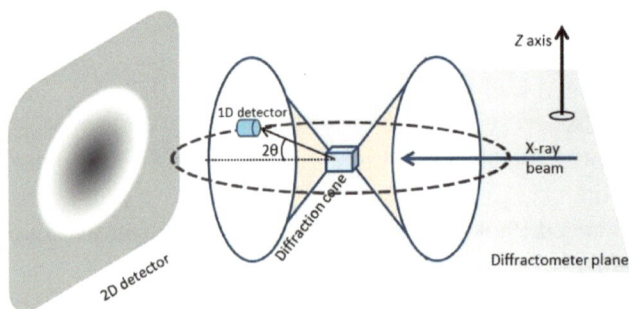

Figure 1.41: Diffraction patterns from a powder sample. Adapted from B. B. He, U. Preckwinkel, and K. L. Smith, Fundamentals of two-dimensional X-ray diffraction (XRD 2). *Adv. X-ray Anal.*, **2000, 43, 273. Copyright: EVISA (2000).**

One such advantage is the measurement of percent crystallinity of a material. Determination of material crystallinity is required both for research and quality control. Scattering from amorphous materials produces a di use intensity ring while polycrystalline samples produce sharp and well-defined rings or spots are seen. The ability to distinguish between amorphous and crystalline is the key in determining percent of crystallinity accurately. Since most crystalline samples have preferred orientation, depending on the sample is oriented it is possible to measure different peak or no peak using conventional diffraction system. On the other hand, sample orientation has no effect on the full circle integrated diffraction measuring done using 2D detector. A 2D XRD can therefore measure percent crystallinity more accurately.

2D wide angle X-ray diffraction patterns of LCs

As mentioned in the introduction section, liquid crystal is an intermediate state between solid and liquid phases. At temperatures above the liquid crystal phase transition temperature (Figure 1.42), they become isotropic liquid, i.e., absence of long-range positional or orientational order within molecules. Since an isotropic state cannot be aligned, its diffraction pattern consists of weak, di use rings (Figure 1.42a). The reason we see any diffraction pattern in the isotropic state is because in classical liquids there exists a short-range positional order. The ring has of radius of 4.5 Å and it mostly appears at 20.5°. It represents the distance between the molecules along their widths.

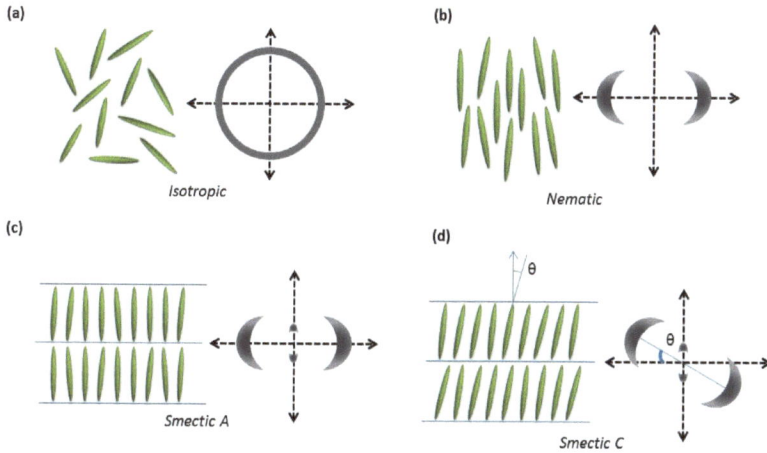

Figure 1.42: Schematic of 2D X-ray diffraction of different types of liquid crystal phases: (a) isotopic, (b) nematic, (c) smectic A, and (d) smectic C.

Nematic liquid crystalline phases have long range orientational order but no positional order. An un- aligned sample of nematic liquid crystal has similar diffraction pattern as an isotropic state. But instead of a diffuse ring, it has a sharper intensity distribution. For an aligned sample of nematic liquid crystal, X-ray diffraction patterns exhibit two sets of diffuse arcs (Figure 1.42b). The diffuse arc at the larger radius (P1, 4.5 Å) represents the distance between molecules along their widths. Under the presence of an external magnetic field, samples with positive diamagnetic anisotropy align parallel to the field and P1 is oriented perpendicularly to the field. While samples with negative diamagnetic anisotropy align perpendicularly to the field with P1 being parallel to the field. The intensity distribution within these arcs represents the extent of alignment within the sample; generally denoted by S.

The diamagnetic anistropy of all liquid crystals with an aromatic ring is positive, and on order of 10^{-7}. The value decreases with the substitution of each aromatic ring by a cyclohexane or other aliphatic group. A negative diamagnetic anistropy is observed for purely cycloaliphatic LCs.

When a smectic phase is cooled down slowly under the presence the external field, two sets of di use peaks are seen in diffraction pattern (Figure 1.42c). The di use peak at small angles condense into sharp quasi- Bragg peaks. The peak intensity distribution at large angles are not very sharp because molecules within the smectic planes are randomly arranged. In case of smectic C phases, the angle between the smectic layers normal and the director (θ) is no longer collinear (Figure 1.42d). This tilt can easily be seen in the diffraction pattern as the di use peaks at smaller and larger angles are no longer orthogonal to each other.

Sample preparation

In general, X-ray scattering measurements of liquid crystal samples are considered more di cult to perform than those of crystalline samples. The following steps should be performed for diffraction measurement of liquid crystal samples:

- The sample should be free of any solvents and absorbed oxygen, because their presence affects the liquid crystalline character of the sample and its thermal response. This can be achieved by performing multiple melting and freezing cycles in a vacuum to get rid of unwanted solvents and gases.
- For performing low resolution measurements, liquid crystal sample can be placed inside a thin-walled glass capillary. The ends of the capillary can be sealed by epoxy in case of volatile samples. The filling process tends to align the liquid crystal molecules along the flow direction.
- For high resolution measurements, the sample is generally confined between two rubbed polymer coated glass coverslips separated by an o-ring as a spacer. The rubbing causes formation of grooves in the polymer film which tends to the align the liquid crystal molecules.
- Aligned samples are necessary for identifying the liquid crystalline phase of the sample. Liquid crystal samples can be aligned by heating above the phase transition temperature and cooling them slowly in the presence of an external electric or magnetic field. A magnetic field is effective for samples with aromatic cores as they have high diamagnetic anisotropy. A common problem in using electric field is internal heating which can interfere with the measurement.

- Sample size should be sufficient to avoid any obstruction to the passage of the incident X-ray beam.
- The sample thickness should be around one absorption length of the X-rays. This allows about 63% of the incident light to pass through and get optimum scattering intensity. For most hydrocarbons the absorption length is approximately 1.5 mm with a copper metal target ($\lambda = 1.5418$ Å). Molybdenum target can be used for getting an even higher energy radiation ($\lambda = 0.71069$ Å).

Data analysis

Identification of the phase of a liquid crystal sample is critical in predicting its physical properties. A simple 2D X-ray diffraction pattern can tell a lot in this regard (Figure 1.42). It is also critical to determine the orientational order of a liquid crystal. This is important to characterize the extent of sample alignment.

For simplicity, the rest of the discussion focuses on nematic liquid crystal phases. In an unaligned sample, there isn't any specific macroscopic order in the system. In the micrometer size domains, molecules are all oriented in a specific direction, called a local director. Because there is no positional order in nematic liquid crystals, this local director varies in space and assumes all possible orientations. For example, in a perfectly aligned sample of nematic liquid crystals, all the local directors will be oriented in the same direction. The specific alignment of molecules in one preferred direction in liquid crystals makes their physical properties such as refractive index, viscosity, diamagnetic susceptibility, directionally dependent.

When a liquid crystal sample is oriented using external fields, local directors preferentially align globally along the field director. This globally preferred direction is referred to as the director and is denoted by unit vector n. The extent of alignment within a liquid crystal sample is typically denoted by the order parameter, S, as defined by

$$S = (3\cos^2\theta - 1)/2$$

where θ is the angle between long axis of molecule and the preferred direction, n.

For isotropic samples, the value of S is zero, and for perfectly aligned samples it is 1. Figure 1.43 shows the structure of a most extensively studied nematic liquid crystal molecule, 4-cyano-4'-pentylbiphenyl, commonly known as 5CB. For preparing a polydomain sample 5CB was placed inside a glass capillary via capillary forces (Figure 1.44). Figure 1.45 shows the 2D X-ray diffraction of the as prepared polydomain sample.

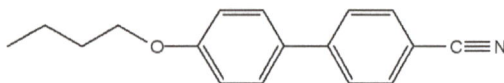

Figure 1.43: Chemical structure of a nematic liquid crystal molecule 4-cyano-4'-pentylbiphenyl (also known as 5CB).

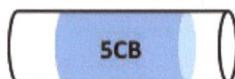

Figure 1.44: Schematic representation of a polydomain liquid crystal samples (5CB) inside a glass capillary.

Figure 1.45: 2D X-ray diffraction of polydomain nematic liquid crystal sample of 5CB. Data was ac- quired using a Rigaku Raxis-IV++ equipped with an incident beam monochromator, pinhole collimation (0.3 mm) and Cu X-ray tube (λ = 1.54 Å). The sample to detector distance was 100 mm.

For preparing monodomain sample, a glass capillary filled with 5CB was heated to 40 °C (i.e., above the nematic-isotropic transition temperature of 5CB, ~35 °C) and then cooled slowly in the presence of magnetic field (1 Testla, Figure 1.46). This gives a uniformly aligned sample with the nematic director n oriented along the magnetic field. Figure 1.47 shows the collected 2D X-ray diffraction measurement of a monodomain 5CB liquid crystal sample using Rigaku Raxis-IV++, and it consists of two diffuse arcs (as mentioned before). Figure 1.48 shows the intensity distribution of a diffuse arc as a function of Θ, and the calculated order parameter value, S, is -0.48.

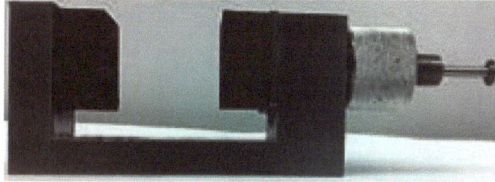

Figure 1.46: Magnetic field setup used to prepare a monodomain sample of 5CB. The glass capillary can just be seen between the sides of the holder.

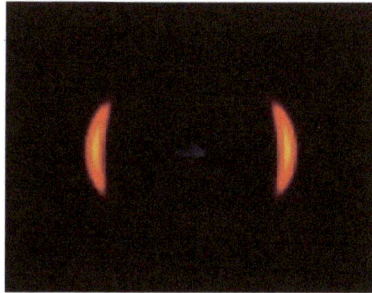

Figure 1.47: 2D X-ray diffraction of polydomain nematic liquid crystal sample of 5CB. Data was acquired using a Rigaku Raxis-IV++ equipped with an incident beam monochromator, pinhole collimation (0.3 mm) and Cu X-ray tube (λ = 1.54 Å). The sample to detector distance was 100 mm.

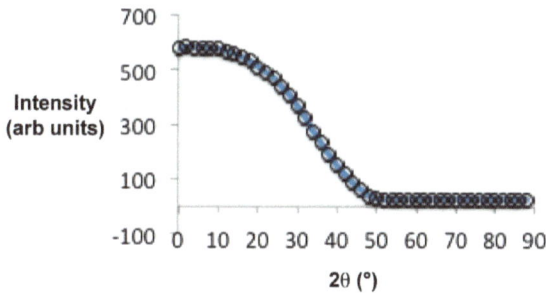

Figure 1.48: Plot of intensity versus 2θ (°) for a 2D X-ray diffraction measurement of the monodomain sample of 5CB.

Refinement of crystallographic disorder in the tetra-fluoroborate Anion

Through the course of our structural characterization of various tetrafluoroborate salts, the complex cation has nominally been the primary subject of interest; however, we observed that the tetrafluoroborate anion (BF_4^-) anions were commonly disordered (13 out of 23 structures investigated). Furthermore, a consideration of the Cambridge Structural Database as of 14th December 2010 yielded 8,370 structures in which the tetrafluoroborate anion is present; of these, 1044 (12.5%) were refined as having some kind of disorder associated with the BF_4^- anion. Several different methods have been reported for the treatment of these disorders, but the majority was refined as a non-crystallographic rotation along the axis of one of the B-F bonds.

Unfortunately, the very property that makes fluoro-anions such good candidates for non-coordinating counter-ions (i.e., weak intermolecular forces) also facilitates the presence of disorder in crystal structures. In other words, the appearance of disorder is intensified with the presence of a weakly coordinating spherical anion (e.g., BF_4^- or PF_6^-) which lack the strong intermolecular interactions needed to keep a regular, repeating anion orientation throughout the crystal lattice. Essentially, these weakly coordinating anions are loosely defined electron-rich spheres. All considered it seems that fluoro-anions, in general, have a propensity to exhibit apparently large atomic displacement parameters (ADP's), and thus, are appropriately refined as having fractional site-occupancies.

Refining disorder

In crystallography the observed atomic displacement parameters are an average of millions of unit cells throughout entire volume of the crystal, and thermally induced motion over the time used for data collection. A disorder of atoms/molecules in a given structure can manifest as at or non-spherical atomic displacement parameters in the crystal structure. Such cases of disorder are usually the result of either thermally induced motion during data collection (i.e., dynamic disorder), or the static disorder of the atoms/molecules throughout the lattice. The latter is defined as the situation in which certain atoms, or groups of atoms, occupy slightly different orientations from molecule to molecule over the large volume (relatively speaking) covered by the crystal lattice. This static displacement of atoms can simulate the effect of thermal vibration on the scattering power of the "average" atom.

Consequently, differentiation between thermal motion and static disorder can be ambiguous, unless data collection is performed at low temperature (which would negate much of the thermal motion observed at room temperature).

In most cases, this disorder is easily resolved as some non-crystallographic symmetry elements acting locally on the weakly coordinating anion. The atomic site occupancies can be refined using the FVAR instruction on the different parts (see PART 1 and PART 2 in Figure 1.49) of the disorder, having a site occupancy factor (s.o.f.) of $\times$ and 1-x, respectively. This is accomplished by replacing 11.000 (on the F-atom lines in the NAME.INS le) with 21.000 or -21.000 for each of the different parts of the disorder. For instance, the "NAME.INS" le would look something like that shown in Figure 1.49. Note that for more heavily disordered structures, i.e., those with more than two disordered parts, the SUMP command can be used to determine the s.o.f. of parts 2, 3, 4, etc. the combined sum of which is set at s.o.f. = 1.0. These are designated in FVAR as the second, third, and fourth terms.

ISOR	$F					
ISOR	$F					
DELU	$F					
SADI	B(1) F(1A)	B(1) F(2A)	B(1) F(3A)	B(1) F(4A)		
SADI	B(1) F(1B)	B(1) F(2B)	B(1) F(3B)	B(1) F(4B)		
SADI	F(1A) F(2A)	F(1A) F(3A)	F(1A) F(4A)	etc.		
SADI	F(1B) F(2B)	F(1B) F(3B)	F(1B) F(4B)	etc.		
FVAR	0.1	0.5 [a]				
B(1)	3	x	y	z	U_{eq}	11.000
PART 1[a]						
F(1A)	6	x^1	y^1	z^1	U_{eq}	21.000
F(2A)	6	x^2	y^2	z^2	U_{eq}	21.000
F(3A)	6	x^3	y^3	z^3	U_{eq}	21.000
F(4A)	6	x^4	y^4	z^4	U_{eq}	21.000
PART 2						
F(1B)	6	x^1	y^1	z^1	U_{eq}	-21.000
F(2B)	6	x^2	y^2	z^2	U_{eq}	-21.000
F(3B)	6	x^3	y^3	z^3	U_{eq}	-21.000
F(4B)	6	x^4	y^4	z^4	U_{eq}	-21.000
PART 0						

Figure 1.49: General layout of the SHELXTL "NAME.INS" le for treatment of disordered tetrafluoroborate. [a]For more than two site occupancies SUMP = 1.0 0.01 1.0 2 1.0 3 1.0 4 is added in addition to the FVAR instruction.

In small molecule refinement, the case will inevitably arise in which some kind of restraints or constraints must be used to achieve convergence of the data. A restraint is any additional information concerning a given structural feature, i.e., limits on the possible values of parameters, may be added into the refinement, thereby increasing the number of refined parameters. For example, aromatic systems are essentially at, so for refinement purposes, a troublesome ring system could be restrained to lie in one plane. Restraints are not exact, i.e., they are tied to a probability distribution, whereas constraints are exact mathematical conditions. Restraints can be regarded as falling into one of several general types:

- Geometric restraints, which relates distances that should be similar.
- Rigid group restraints.
- Anti-bumping restraints.
- Linked parameter restraints.
- Similarity restraints.
- ADP restraints (Figure 1.50).
- Sum and average restraints.
- Origin fixing and shift limiting restraints.
- Those imposed upon atomic displacement parameters.

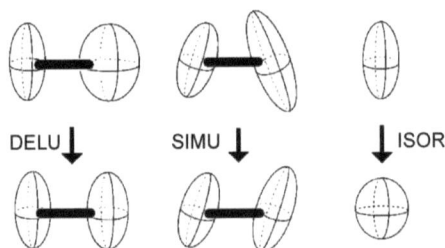

Figure 1.50: Consequence of the anisotropic displacement parameter (ADP) restraints DELU, SIMU, and ISOR on the shape and directionality of atomic displacement parameters. Adapted from P. Maüller, *Crystal Structure Refinement, A Crystallographer's Guide to SHELXL*, Oxford University Press, UK (2006).

Geometric restraints

- SADI - similar distance restraints for named pairs of atoms.
- DFIX - defined distance restraint between covalently bonded atoms.
- DANG - defined non-bonding distance restraints, e.g., between F atoms belonging to the same PART of a disordered BF_4^-.

- FLAT - restrains group of atoms to lie in a plane.

Anisotropic displacement parameter restraints
- DELU - rigid bond restraints (Figure 1.50).
- SIMU - similar ADP restraints on corresponding U_{ij} components to be approximately equal for atoms in close proximity (Figure 1.50).
- ISOR - treat named anisotropic atoms to have approximately isotropic behavior (Figure 1.50).

Constraints (different than "restraints")
- EADP - equivalent atomic displacement parameters.
- AFIX - fitted group; e.g., AFIX 66 would fit the next six atoms into a regular hexagon.
- HFIX - places H atoms in geometrically ideal positions, e.g., HFIX 123 would place two sets of methyl H atoms disordered over two sites, 180° from each other.

Classes of disorder for the tetrafluoroborate anion

Rotation about a non-crystallographic axis along a B-F bond

The most common case of disorder is a rotation about an axis, the simplest of which involves a non- crystallographic symmetry related rotation axis about the vector made by one of the B-F bonds; this operation leads to three of the four F-atoms having two site occupancies (Figure 1.51). This disorder is also seen for ^{t}Bu and CF_3 groups, and due to the C_3 symmetry of the $C(CH_3)_3$, CF_3 and BF_3 moieties actually results in a near C2 rotation.

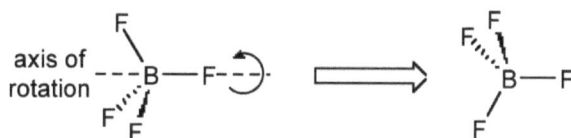

Figure 1.51: Schematic representation of the rotational relationship between two disordered orientations of the BF$_4^-$ anion.

In a typical example, the BF_4^- anion present in the crystal structure of [H(Mes-dpa)]BF$_4$ (Figure 1.52a, where R = 2,4,6-Me$_3$C$_6$H$_2$) was found to have a 75:25 site occupancy disorder for three of the four fluorine atoms (Figure 1.53). The disorder is a rotation about the axis of the B(1)-F(1) bond. For initial refinement cycles, similar distance restraints (SADI) were placed on all B-F and F-F distances, in addition to similar ADP restraints (SIMU) and rigid bond restraints (DELU) for all F atoms. Restraints were lifted for final refinement cycles. A similar disorder refinement was required for [H(2-^{i}PrPh-dpa)]BF$_4$ (45:55), while refinement of the disorder in [Cu(2-^{i}PrPh-dpa)(styrene)]BF$_4$ (65:35) was performed with only SADI and DELU restraints were lifted in final refinement cycles.

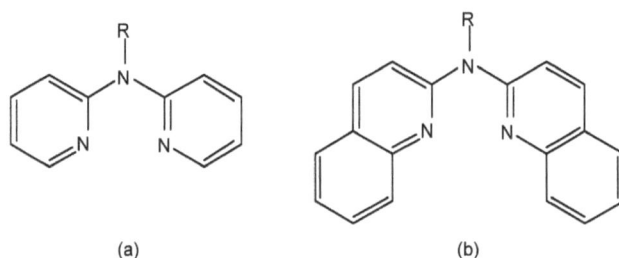

(a)   (b)

Figure 1.52: Structures of (a) substituted *bis*(2-pyridyl)amines (R-dpa) and (b) substituted *bis*(2- quinolyl)amines [R-N(quin)₂] ligands.

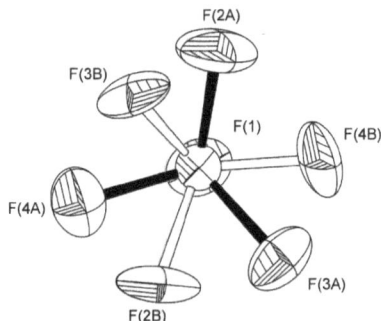

Figure 1.53: Structure for the BF$_4^-$ anion in compound [H(Mes-dpa)]BF$_4$ with both parts of the disorder present. Thermal ellipsoids are shown at the 20% level. Adapted from J. J. Allen, Steric considerations in copper(II)-olefin complexes incorporating substituted *bis*-2-pyrol)amines. PhD Thesis, Rice University (2011).

In the complex [Ag(H-dpa)(styrene)]BF$_4$ use of the free variable (FVAR) led to refinement of disordered fluorine atoms F(2A)-F(4A) and F(2B)-F(4B) as having a 75:25 site-occupancy disorder (Figure 1.54). For initial refinement cycles, all B-F bond lengths were given similar distance restraints (SADI). Similar distance restraints (SADI) were also placed on F···F distances for each part, i.e., F(2A)···F(3A) = F(2B)···F(3B), etc. Additionally, similar ADP restraints (SIMU) and rigid bond restraints (DELU) were placed on all F atoms. All restraints, with the exception of SIMU, were lifted for final refinement cycles.

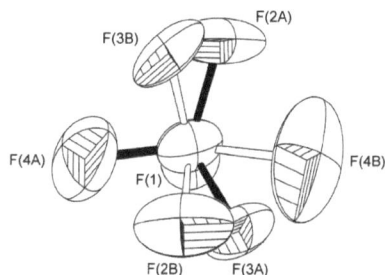

Figure 1.54: Structure of the disordered BF₄⁻ anion in [Ag(H-dpa)(styrene)]BF4 viewed down the axis of disorder. Thermal ellipsoids are shown at the 30% probability level. Adapted from J. J. Allen, Steric considerations in copper(II)-olefin complexes incorporating substituted *bis*-2-pyrol)amines. PhD Thesis, Rice University (2011).

Rotation about a non-crystallographic axis not along a B-F bond

The second type of disorder is closely related to the first, with the only difference being that the rotational axis is tilted slightly off the B-F bond vector, resulting in all four F-atoms having two site occupancies (Figure 1.55). Tilt angles range from 6.5° to 42°.

The disordered BF⁻ anion present in the crystal structure of [Cu(Ph-dpa)(styrene)]BF₄ was refined having fractional site occupancies for all four fluorine atoms about a rotation slightly tilted off the B(1)-F(2A) bond. However, it should be noted that while the $U_{(eq)}$ values determined for the data collected at low temperature data is roughly half that of that found at room temperature, as is evident by the sizes and shapes of fluorine atoms in Figure 1.56, the site occupancies were refined to 50:50 in each case, and there was no resolution in the disorder.

An extreme example of rotation off-axis is observed where refinement of more than two site occupancies (Figure 1.57) with as many as thirteen different fluorine atom locations on only one boron atom.

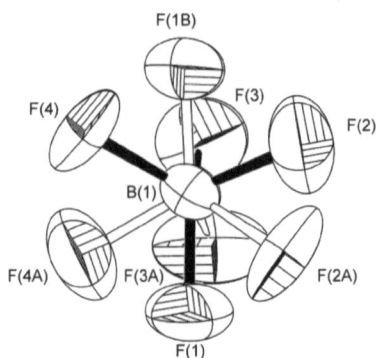

Figure 1.55: Molecular structure for the anion in [Cu(H-dpa)(*cis*-3-octene)]BF₄ with both parts of the disordered BF₄⁻ present. Thermal ellipsoids are shown at the 20% level. Adapted from J. J. Allen, Steric considerations in copper(II)-olefin complexes incorporating substituted *bis*-2-pyrol)amines. PhD Thesis, Rice University (2011).

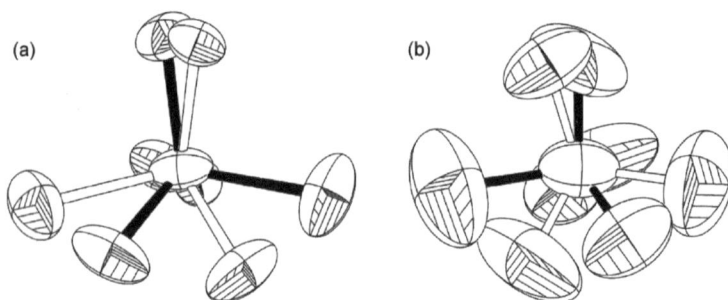

Figure 1.56: Comparison of the atomic displacement parameters observed in the disordered BF₄⁻ anion from [Cu(Ph-dpa)(styrene)]BF₄ at data collection temperature (a) T = 213 K and (b) T = 298 K. Thermal ellipsoids are set at the 25% level. Adapted from J. J. Allen, Steric considerations in copper(II)-olefin complexes incorporating substituted *bis*-2-pyrol)amines. PhD Thesis, Rice University (2011).

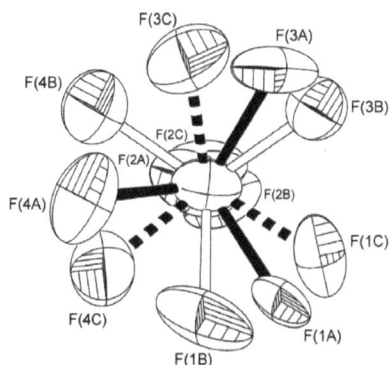

Figure 1.57: Structure for the tetrafluoroborate anion with twelve fluorine atom locations. Data from S. Martinez-Vargas, R. Toscano, and J. Valdez-Martinez, (3-Carboxypyridine-2-carboxylato-κ^2N,O)(4'-phenyl-2,2':6',2"-terpyridine-κ^3N,N',N")copper(II) tetrafluoridoborate. *Acta Cryst.*, 2007, E63, m1975.

Constrained rotation about a non-crystallographic axis not along a B-F bond

Although a wide range of tilt angles are possible, in some systems the angle is constrained by the presence of hydrogen bonding. For example, the BF_4^- anion present in [Cu(Mes-dpa)(μ-OH)(H$_2$O)][BF$_4$] was found to have a 60:40 site occupancy disorder of the four fluorine atoms, and while the disorder is a C_2-rotation slightly tilted of the axis of the B-F bond, the angle is restricted by the presence of two B-F···O interactions for one of the isomers (Figure 1.58).

An example that does adhere to global symmetry elements is seen in the BF_4^- anion of [Cu{2,6-^{i}Pr$_2$C$_6$H$_3$N(quin)$_2$}$_2$]BF$_4$.MeOH (Figure 1.59), which exhibits a hydrogen-bonding interaction with a disor- dered methanol solvent molecule. The structure of R-N(quin)$_2$ is shown in Figure 1.52b. By crystallographic symmetry, the carbon atom from methanol and the boron atom from the BF_4^- anion lie on a C_2-axis. Fluorine atoms [F(1)-F(4)], the methanol oxygen atom, and the hydrogen atoms attached to methanol O(1S) and C(1S) atoms were refined as having 50:50 site occupancy disorder (Figure 1.59).

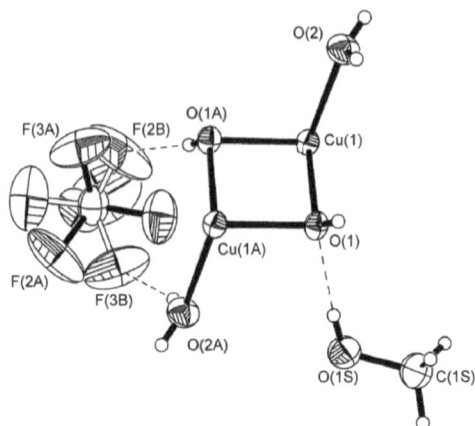

Figure 1.58: Structure of the disordered BF₄⁻ in [Cu(Mes-dpa)(μ-OH)(H₂O)]₂[BF₄]₂ showing interaction with bridging hydroxide and terminal water ligands. Thermal ellipsoids are shown at the 20% level. Adapted from J. J. Allen, Steric considerations in copper(II)-olefin complexes incorporating substituted *bis*-2-pyrol)amines. PhD Thesis, Rice University (2011).

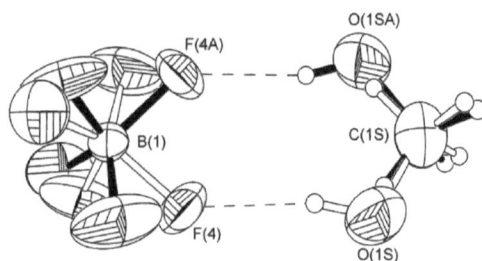

Figure 1.59: H-bonding interaction in [Cu{2,6-^{i}Pr₂C₆H₃N(quin)₂}₂]BF₄.MeOH between anion and solvent of crystallization, both disordered about a crystallographic C₂-rotation axis running through the B(1)···C(1S) vector. Adapted from J. J. Allen, Steric considerations in copper(II)-olefin complexes incorporating substituted *bis*-2-pyrol)amines. PhD Thesis, Rice University (2011).

Non crystallographic inversion center at the boron atom

Multiple disorders can be observed with a single crystal unit cell. For example, the two BF₄⁻ anions in [Cu(Mes-dpa)(styrene)]BF₄ both exhibited 50:50 site occupancy disorders, the first is a C₂-rotation tilted off one of the B-F bonds, while the second is disordered about an inversion centered on the boron atom. Refinement of the latter was carried out similarly to the aforementioned cases, with the exception that fixed distance restraints for non-bonded atoms (DANG) were left in place for the disordered fluorine atoms attached to B (Figure 1.60).

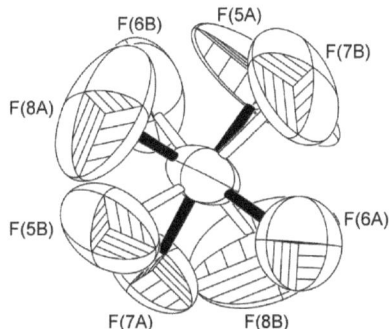

Figure 1.60: Structure for the disordered BF$_4$⁻ anion due to a NCS-inversion center, in compound [Cu(Mes-dpa)(styrene)]BF$_4$ with both parts of the disorders present. Thermal ellipsoids are shown at the 20% level. Adapted from J. J. Allen, Steric considerations in copper(II)-olefin complexes incorporating substituted *bis*-2-pyrol)amines. PhD Thesis, Rice University (2011).

Disorder on a crystallographic mirror plane

Another instance in which the BF$_4$⁻ anion is disordered about a crystallographic symmetry element is that of [Cu(H-dpa)(1,5-cyclooctadiene)]BF$_4$. In this instance fluorine atoms F(1) through F(4) are present in the asymmetric unit of the complex. Disordered atoms F(1A)-F(4A) were refined with 50% site occupancies B(1) lies on a mirror plane (Figure 1.61). For initial refinement cycles, similar distance restraints (SADI) were placed on all B-F and F-F distances, in addition to similar ADP restraints (SIMU) and rigid bond restraints (DELU) for all F atoms. Restraints were lifted for final refinement cycles, in which the boron atom lies on a crystallographic mirror plane, and all four fluorine atoms are reflected across.

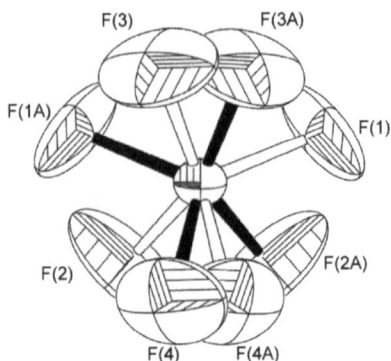

Figure 1.61: Molecular structure for the anion in [Cu(H-dpa)(1,5-cyclooctadiene)]BF₄ with both parts of the disordered BF₄⁻ present. For clarity, thermal ellipsoids are shown at the 20% level. Adapted from J. J. Allen, Steric considerations in copper(II)-olefin complexes incorporating substituted *bis*-2-pyrol)amines. PhD Thesis, Rice University (2011).

Disorder on a non-crystallographic mirror plane

It has been observed that the BF_4^- anion can exhibit site occupancy disorder of the boron atom and one of the fluorine atoms across an NCS mirror plane defined by the plane of the other three fluorine atoms (Figure 1.62) modeling the entire anion as disordered (including the boron atom).

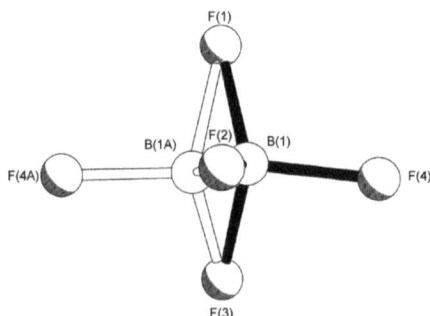

Figure 1.62: Disordered anion across the plane of three fluorine atoms. Data from J. T. Mague and S. W. Hawbaker, 2-pyridylbis(diphenylphosphino)methane chemistry. Synthesis and structures of [Cu(μ-η²:η¹(Ph₂P)₂-CHC₅H₄N)(THF)]₂(BF₄)₂ and [Ni(Ph₂PCH₂C₅H₄N)₂]-[NiCl₄]·0.85CH₂Cl₂and [Ni(Ph₂PCH₂C₅H₄N)₂]-[NiCl₄]·0.85CH₂Cl₂. *J. Chem. Cryst.*, 1997, 27, 603.

Disorder of the boron atom core

The extreme case of a disorder involves refinement of the entire anion, with all boron and all fluorine atoms occupying more than two sites (Figure 1.63). In fact, some disorders of the latter types must be refined isotropically, or as a last resort, not at all, to prevent one or more atoms from turning non-positive definite.

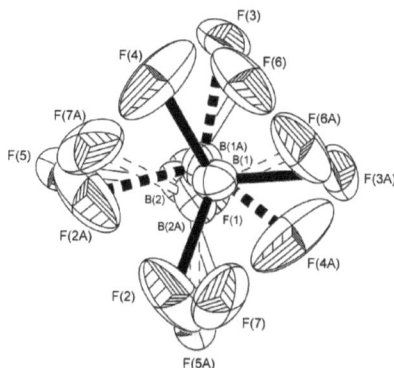

Figure 7.16: An example of a structure of a highly disordered BF$_4^-$ anion refined with four site occupancies for all boron and fluorine atoms. Data from P. Szklarz, M. Flowczarek, G. Bator, T. Lis, K. Gatner, and R. Jakubas, Crystal structure, properties and phase transitions of morpholinium tetrafluoroborate [C$_4$H$_{10}$NO][BF$_4$]. *J. Mol. Struct.*, 2009, 929, 48.

Bibliography

J. J. Allen, Steric considerations in copper(II)-olefin complexes incorporating substituted *bis*-2-pyrol)amines. PhD Thesis, Rice University (2011).

J. J. Allen and A. R. Barron, Synthesis and structural characterization of [Ag(H-dpa)(η^2-styrene)]BF$_4$: comparing silver and copper for olefin binding. *J. Chem. Cryst.*, 2009, **39**, 935.

J. J. Allen and A. R. Barron, Olefin coordination in copper(I) complexes of *bis*(2-pyridyl)amine. *Dalton Trans.*, 2009, 878.

J. J. Allen, C. E. Hamilton, and A. R. Barron, Synthesis and characterization of aryl-substituted *bis*(2-pyridyl)amines and their copper olefin complexes: investigation of remote steric control over olefin binding,. *Dalton Trans.*, 2010, 11451.

J. B. Brady and R. M. Newton, New uses for powder X-ray diffraction experiments in the undergraduate curriculum. *J. Geol. Educ.*, 1995, **43**, 466.

L. Brugemann and E. K. E. Gerndt, Detectors for X-ray diffraction and scattering: a user's overview. *Nucl. Instrum. Meth. A*, 2004, **531**, 292.

H. W. Chiu, C. N. Chervin, and S. M. Kauzlarich, Phase changes in Ge nanoparticles. *Chem. Mater.*, 2005, **17**, 4858.

W. Clegg, *Crystal Structure Determination*, Oxford University Press, Oxford (1998).

B. D. Cullity and S. R. Stock. *Elements of X-ray Diffraction*, 3rd edn., Prentice Hall, New Jersey (2001).

P. G. de Gennes, *Physics of Liquid Crystals*, Oxford University Press, New York (1995).

R. D. Deslattes, E.G. Kessler, Jr., P. Indelicato, L. de Billy, E. Lindroth, and J. Anton, X-ray transition energies: new approach to a comprehensive evaluation. *Rev. Mod. Phys.*, 2003, **75**, 35.

E. A. V. Ebsworth, D. W. H. Rankin, and S. Cradock, *Structural Methods in Inorganic Chemistry*, 2nd edn., CRC Bress, Boca Raton (1991).

J. P. Glusker, M. Lewis, and M. Rossi, *Crystal Structure Analysis for Chemists and Biologists*, VCH, New York (1994).

B. B. He, U. Preckwinkel, and K. L. Smith, Fundamentals of two-dimensional X-ray diffraction (XRD 2). *Adv. X-ray Anal.*, 2000, **43**, 273.

F. L. Hirshfield, Can X-ray data distinguish bonding effects from vibrational smearing?. *Acta Cryst.*, 1976, **A32**, 239.

K. T. Howard, G. K. Benedix, P. A. Bland, and G. Cressey, Modal mineralogy of CM2 chondrites by X-ray diffraction (PSD-XRD). Part 1: Total phyllosilicate abundance and the degree of aqueous alteration. *Geochim. Cosmochim. Acta*, 2009, **73**, 4576.

H. D. Jakubke and H. Jeschkeit, *Concise Encyclopedia Chemistry*, Walter de Gruyter, Berlin (1993).

B. Lavina, P. Dera, and R. T. Downs, Modern X-ray diffraction methods in mineralogy and geosciences. *Rev. Mineral. Geochem.*, 2014, **78**, 1.

J. T. Mague and S. W. Hawbaker, 2-pyridylbis(diphenylphosphino)methane chemistry. Synthesis and structures of $[Cu(\mu-\eta^2:\eta^1(Ph_2P)_2-CHC_5H_4N)(THF)]_2(BF_4)_2$ and $[Ni(Ph_2PCH_2C_5H_4N)_2]-[NiCl_4]\cdot0.85CH_2Cl_2$ and $[Ni(Ph_2PCH_2C_5H_4N)_2]-[NiCl_4]\cdot0.85CH_2Cl_2$. *J. Chem. Cryst.*, 1997, **27**, 603.

S. Martinez-Vargas, R. Toscano, and J. Valdez-Martinez, (3-Carboxypyridine-2-carboxylato-κ^2N,O)(4'-phenyl-2,2':6',2"-terpyridine-κ^3N,N',N")copper(II) tetrafluoridoborate. *Acta Cryst.*, 2007, **E63**, m1975.

P. Maüller, Crystal Structure Refinement, A Crystallographer's Guide to SHELXL, Oxford University Press, UK (2006).

D. M. Moore and R. C. Reynolds. *X-Ray Diffraction and the Identification and Analysis of Clay Minerals*. 2nd edn., Oxford University Press (1997).

K. Mukherjee, Ab-initio crystal structure determination from X-ray powder diffraction data. *J. Indian Inst. Sci.*, 2007, **87**, 221.

D. F. Mullica, S. L. Gibson, E. L. Sappen field, C. C. Lin, and D. H. Leschnitzer, Inorg. Chim. Acta, 1990, 177, 89.

N. S. Murthy and H. Minor, General procedure for evaluating amorphous scattering and crystallinity from X-ray diffraction scans of semicrystalline polymers. *Polymer*, 1990, **31**, 996.

L. Ooi, *Principles of X-ray Crystallography*, Oxford University Press, Oxford (2010).

P. Ptacek, T. Opravil, F. Soukal, J. Tkacz, J. Masilko, and E. Bartonickova, The field of solid solutions in ternary system of synthetic apatite-type alkaline earth element-yttrium-silicate oxybritholite phases of the composition: $AEE_{\delta}Y_{10-\delta}[SiO_4]_6O_{3-0.5\delta}$, where AEE=Ca, Sr and Ba. *Ceram. Int.*, 2016, **42**, 6154.

P. Szklarz, M. Flowczarek, G. Bator, T. Lis, K. Gatner, and R. Jakubas, Crystal structure, properties and phase transitions of morpholinium tetrafluoroborate $[C_4H_{10}NO][BF_4]$. *J. Mol. Struct.*, 2009, **929**, 48.

M. Trachsel, U. Eggenberger, M. Grosjean, A. Blass, and M. Sturm, Mineralogy-based quantitative precipitation and temperature reconstructions from annually laminated lake sediments (Swiss Alps) since AD 1580. *Geophys. Res. Lett.*, 2008, **35**, L13707.

F. J. Turner, Preferred Orientation of Calcite and Dolomite in Experimentally Deformed Marbles, Office of Naval Research (1952).

USGS Open-File Report 01-041. https://pubs.usgs.gov/of/2001/of01041/htmldocs/flow/flow.pdf.

M. von Laue, Concerning the detection of X-ray interferences, Nobel Prize. Nobel Foundation (1915).

X-Ray Powder Diffraction. U.S. Geological Survey. Web. https://pubs.usgs.gov/info/diffraction/html/.

W. Zhu, C. Lu, F. Chang, and S. Kuo, Supramolecular ionic strength-modulating microstructures and properties of nacre-like biomimetic nanocomposites containing high loading clay. *RSC Adv.*, 2012, **2**, 6295.

Chapter 2: Low Energy Electron Diffraction

Gladys A. López-Silva and Andrew R. Barron

Introduction

Low energy electron diffraction (LEED) is a very powerful technique that allows for the characterization of the surface of materials. Its high surface sensitivity is due to the use of electrons with energies between 20 - 200 eV, which have wavelengths equal to 2.7 - 0.87 Å (comparable to the atomic spacing). Therefore, the electrons can be elastically scattered easily by the atoms in the first few layers of the sample. Its features, such as little penetration of low energy electrons have positioned it as one of the most common techniques in surface science for the determination of the symmetry of the unit cell (qualitative analysis) and the position of the atoms in the crystal surface (quantitative analysis).

History

In 1924 Louis de Brogile (Figure 2.1) postulated that all forms of matter, such as electrons, have a wave-particle nature.

Figure 2.1: French physicist Louis Victor Pierre Raymond de Broglie, 7th duc de Broglie (1892 - 1987).

Three years later after this postulate, the American physicists Clinton J. Davisson and Lester H. Germer (Figure 2.2) proved experimentally the wave nature of electrons at Bell Labs in New York. At that time, they were investigating the distribution-in-angle of the elastically scattered electrons (electrons that have suffered no loss of kinetic energy) from the (111) face of a polycrystalline nickel, material composed of many randomly oriented crystals.

Figure 2.2: American physicists Clinton Davisson (left) (1881 - 1958) and Lester Germer (right) (1896 - 1971) in their laboratory, where they proved that electrons could act like waves in 1927.

The experiment consisted of a beam of electrons from a heated tungsten lament directed against the polycrystalline nickel and an electron detector, which was mounted on an arc to observe the electrons at different angles. During the experiment, air entered in the vacuum chamber where the nickel was, producing an oxide layer on its surface. Davisson and Clinton reduced the nickel by heating it at high temperature. They did not realize that the thermal treatment changed the polycrystalline nickel to a nearly monocrystalline nickel, material composed of many oriented crystals. When they repeated the experiment, it was a great surprise that the distribution-in-angle of the scattered electrons manifested sharp peaks at certain angles. They soon realized that these peaks were interference patterns, and, in analogy to X-ray diffraction, the arrangement of atoms and not the structure of the atoms was responsible for the pattern of the scattered electrons.

The results of Davisson and Germer were soon corroborated by George Paget Thomson (Figure 2.3). In 1937, both Davisson and Thomson were awarded with the Nobel Prize in Physics for their experimental discovery of the electron diffraction by crystals. It is noteworthy that 31 years after Thomson's father J. J. Thomson received the Nobel Prize in Physics for showing that the electron is a particle, his son showed that it is also a wave.

Figure 2.3: English physicists Sir George Paget Thomson, FRS (1892 - 1975) who received the Nobel Prize in Physics with Clinton Davisson (Figure 2.2, left) for discovery of the wave properties of the electron by electron diffraction.

Although the discovery of low-energy electron diffraction was in 1927, it became popular in the early 1960's, when the advances in electronics and ultrahigh vacuum technology made possible the commercial availability of LEED instruments. At the beginning, this technique was only used for qualitative characterization of surface ordering. Years later, the impact of computational technologies allowed the use of LEED for quantitative analysis of the position of atoms within a surface. This information is hidden in the energetic dependence of the diffraction spot intensities, which can be used to construct a LEED I-V curve.

Principles and diffraction patterns

Electrons can be considered as a stream of waves that hit a surface and are diffracted by regions with high electron density (the atoms). The electrons in the range of 20 to 200 eV can penetrate the sample for about 10 Å without losing energy. Because of this reason, LEED is especially sensitive to surfaces, unlike X-ray diffraction, which gives information about the bulk-structure of a crystal due to its larger mean free path (around micrometers). Table 2.1 compares general aspects of both techniques.

Like X-ray diffraction, electron diffraction also follows the Bragg's law, see Figure 2.4, where λ is the wavelength, a is the atomic spacing, d is the spacing of the crystal layers, θ is the angle between the incident beam and the reflected beam, and n is an integer. For constructive interference between two waves, the path length difference (i.e., a sinθ in electron diffraction and 2d sinθ for X-ray diffraction, see Figure 2.4) must be an integral multiple of the wavelength.

Low energy electron diffraction	X-ray diffraction
Surface structure determination (high surface sensitivity)	Bulk structures determination
Sample single crystal	Sample single-crystal or polycrystalline
Sample must have an oriented surface, sensitive to impurities	Surface impurities not important
Experiment in ultra-high vacuum	Experiment usually at atmospheric pressure
Experiment done mostly at constant incidence angle and variable wavelength (electron energy)	Constant wavelength and variable incidence angle
Diffraction pattern consists of beams visible at almost all energies	Diffraction pattern consists of beams flashing out at specific wavelengths and angles

Table 2.1: Comparison between low energy electron diffraction and X-ray diffraction.

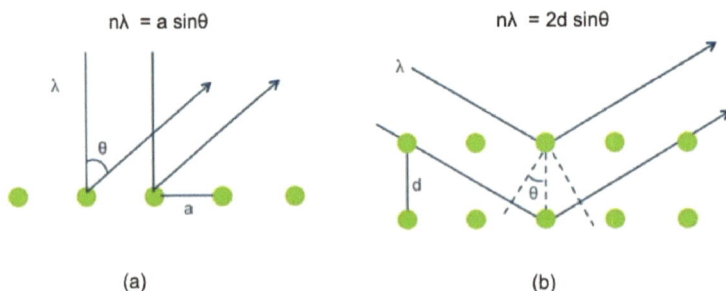

Figure 2.4: Representation of (a) electron and (b) X-ray diffraction.

In LEED, the diffracted beams impact on a fluorescent screen and form a pattern of light spots (Figure 2.5a), which is a to-scale version of the reciprocal lattice of the unit cell. The reciprocal lattice is a set of imaginary points, where the direction of a vector from one point to another point is equal to the direction of a normal to one plane of atoms in the unit cell (real space). For example, an electron beam penetrates a few 2D-atomic layers, Figure 2.6a, so the reciprocal lattice seen by LEED consists of continues rods and discrete points per atomic layer, see Figure 2.6b. In this way, LEED patterns can give information about the size and shape of the real space unit cell, but nothing about the positions of the atoms. To gain this information about atomic positions, analysis of the spot intensities is required.

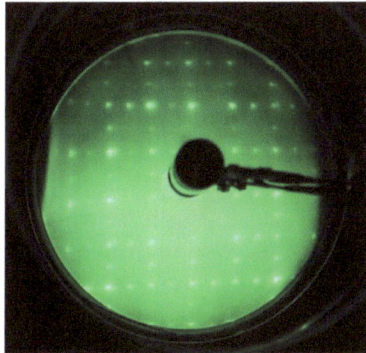

Figure 2.5: LEED pattern of a Si(100) reconstructed surface.

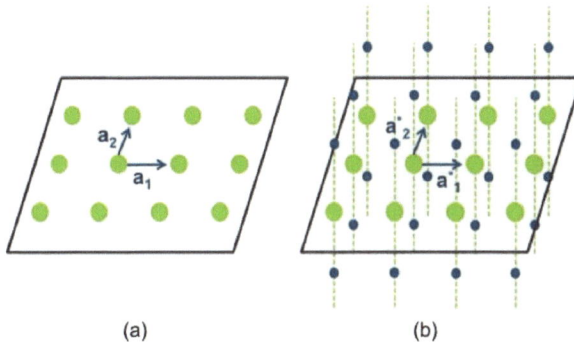

Figure 2.6: (a) 2D atomic layer (real space), and its (b) reciprocal lattice.

Thanks to the hemispheric geometry of the green screen of LEED, we can observe the reciprocal lattice without distortion. It is important to take into account that the separation of the points in the reciprocal lattice and the real interplanar distance are inversely proportional, which means that if the atoms are more widely spaced, the spots in the pattern get closer and vice versa. In the case of superlattices, a periodic structure composed of layers of two materials, new points arise in addition to the original diffraction pattern.

LEED experimental equipment

The typical diagram of a LEED system is shown in Figure 2.7. This system sends an electron beam to the surface of the sample, which comes from an electron gun behind a transparent hemispherical fluorescent screen. The electron gun consists of a heated cathode and a set of focusing lenses which send electrons at low energies. The electrons collide with the sample and diffract in different directions depending on the surface. Once diffracted, they are directed to the fluorescent screen. Before colliding with the screen, they must pass through four different grids (known as retarding grids), which contain a central hole through which the electron gun is inserted. The first grid is the nearest one to the sample and is connected to earth ground. A negative potential is applied to the second and third grids, which act as suppressor grids, given that they repel all electrons coming from non-elastic diffractions. These grids perform as filters, which only allow the highest energy electrons to pass through; the electrons with the lowest energies are blocked in order to prevent a bad resolution image. The fourth grid protects the phosphor screen, which possesses positive charge from the negative grids. The remaining electrons collide with the luminescent screen, creating a phosphor glow (left side of Figure 2.7), where the light intensity depends on the electron intensity.

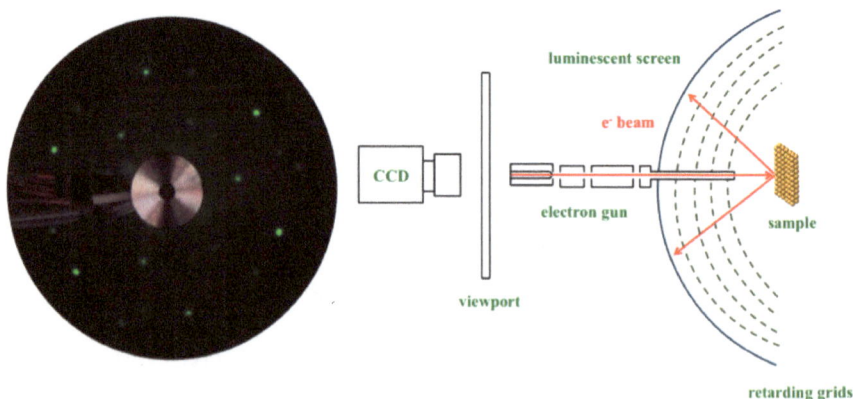

Figure 2.7: Schematic diagram of a typical LEED instrument and an example of the LEED pattern view by the CCD camera. Adapted from L. Meng, Y. Wang, L. Zhang, S. Du, R. Wu, L. Li, Y. Zhang, G. Li, H. Zhou, W. Hofer, H. Gao, Buckled silicene formation on Ir(111). *Nano Lett.*, 2013, 13, 685. Copyright: American Chemical Society (2013).

For conventional systems of LEED, it is necessary a method of data acquisition. In the past, the general method for analyzing the diffraction pattern was to manually take several dozen pictures. After the development of computers, the photographs were scanned and digitalized for further analysis through computational software. Years later, the use of the charge coupled device (CCD) camera was incorporated, allowing rapid acquisition, the possibility to average frames during the acquisition in order to improve the signal, the immediate digitalization and channeling of LEED pattern. In the case of the IV curves, the intensities of the points are extracted making use of special algorithms. Figure 2.8 shows a commercial LEED spectrometer with the CCD camera, which has to be in an ultra-high vacuum vessel.

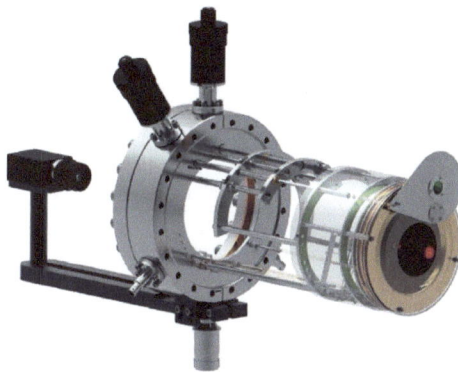

Figure 2.8: Commercial LEED Spectrometer (OCI Vacuum Micro engineering Inc).

LEED applications

Study of adsorbates on the surface and disorder layers

One of the principal applications of LEED is the study of adsorbates on catalysts, due to its high surface sensitivity. In order to illustrate the application of LEED in the study of adsorbates. As an example, Figure 2.9a shows the surface of Cu(100) single crystal, the pristine material. This surface was cleaned carefully by various cycles of sputtering with ions of argon, followed by annealing. The LEED patter of Cu(100) presents four well-defined spots corresponding to its cubic unit cell.

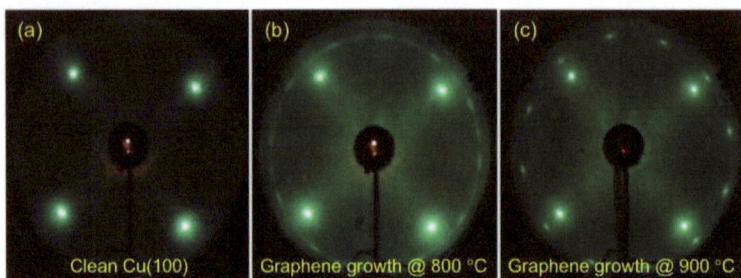

Figure 2.9: LEED patterns of (a) the clean Cu(100) surface, (b) the Cu(100) surface following graphene growth at 800 °C, and (c) the Cu(100) surface following graphene growth at 900 °C. Adapted from Z. Robinson, E. Ong, T. Mowll, P. Tyagi, D. Gaskill, H. Geisler, C. Ventrice, Influence of chemisorbed oxygen on the growth of graphene on Cu(100) by chemical vapor deposition. *J. Phys. Chem. C*, 2013, 117, 23919. Copyright: American Chemical Society (2013).

Figure 2.9b shows the LEED pattern after the growth of graphene on the surface of Cu(100) at 800 °C, we can observe the four spots that correspond to the surface of C (100) and a ring just outside these spots, which correspond to the domains of graphene with four different primary rotational alignments with respect to the Cu(100) substrate lattice, see Figure 2.10. When increasing the temperature of growth of graphene to 900 °C, we can observe a ring of twelve spots (as seen in Figure 2.9c), which indicates that the graphene has a much higher degree of rotational order. Only two domains are observed with an alignment of one of the lattice vectors to one of the Cu(100) surface lattice vectors, given that graphene has a hexagonal geometry, so that only one vector can coincide with the cubic lattice of Cu(100).

One possible explanation for the twelve spots observed at 900 °C is that when the temperature of all domains is increased the four different domains observed at 800 °C, may possess enough energy to adopt the two orientations in which the vectors align with the surface lattice vector of Cu(100). In addition, at 900 °C, a decrease in the size and intensity of the Cu(100) spots is observed, indicating a larger coverage of the copper surface by the domains of graphene.

When the oxygen is chemisorbed on the surface of Cu (100), the new spots correspond to oxygen, Figure 2.11a. Once graphene is allowed to grow on the surface with oxygen at 900 °C, the LEED pattern turns out different: the twelve spots corresponding to graphene domains are not observed due to nucleation of graphene domains in the presence of oxygen in multiple orientations, Figure 2.11b.

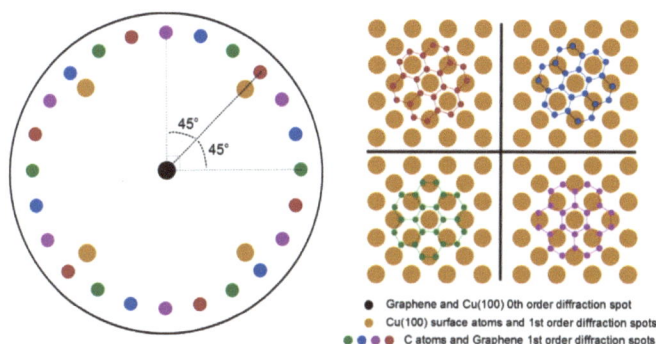

Figure 2.10: Simulated LEED image for graphene domains with four different rotational orientations with respect to the Cu (100) surface. Adapted from Z. Robinson, E. Ong, T. Mowll, P. Tyagi, D. Gaskill, H. Geisler, C. Ventrice, Influence of chemisorbed oxygen on the growth of graphene on Cu(100) by chemical vapor deposition. *J. Phys. Chem. C*, 2013, 117, 23919. Copyright: American Chemical Society (2013).

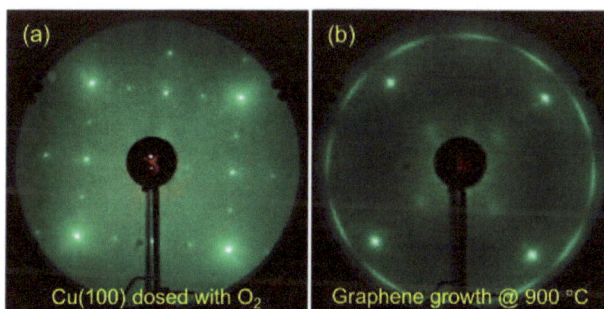

Figure 2.11: LEED patterns of (a) the clean Cu(100) surface dosed with oxygen, (b) the oxygen predosed Cu(100) surface following graphene growth at 900 °C. Adapted from Z. Robinson, E. Ong, T. Mowll, P. Tyagi, D. Gaskill, H. Geisler, C. Ventrice, Influence of chemisorbed oxygen on the growth of graphene on Cu(100) by chemical vapor deposition. *J. Phys. Chem. C*, 2013, 117, 23919. Copyright: American Chemical Society (2013).

A way to study the disorder of the adsorbed layers is through the LEED IV curves, see Figure 2.12. In this case, the intensities are in relation to the angle of the electron beam. The spectrum of Cu(100) with only four sharp peaks shows a very organized surface. In the case of the graphene sample growth over the copper surface, twelve peaks are shown, which correspond to the main twelve spots of the LEED pattern. These peaks are sharp, which indicate a high level of order. For the case of the sample of graphene growth over copper with oxygen, the twelve peaks widen, which is an effect of the increase of disorder in the layers.

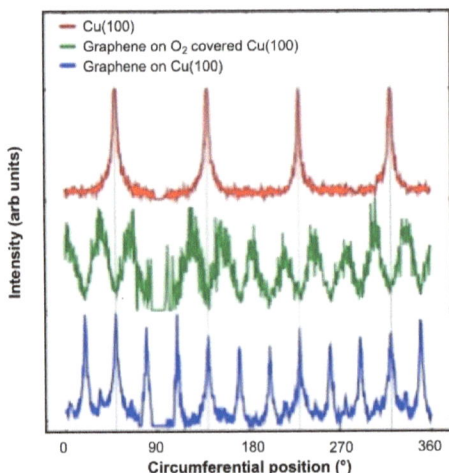

Figure 2.12: LEED-IV using angles for the clean Cu(100) surface (red), graphene grown on the oxygen reconstructed surface (green), and graphene grown on the clean Cu(100) surface (blue). Adapted from Z. Robinson, E. Ong, T. Mowll, P. Tyagi, D. Gaskill, H. Geisler, C. Ventrice, Influence of chemisorbed oxygen on the growth of graphene on Cu(100) by chemical vapor deposition. *J. Phys. Chem. C***, 2013, 117, 23919. Copyright: American Chemical Society (2013).**

Structure determination

As previously mentioned, LEED IV curves may give us exact information about the position of the atoms in a crystal. These curves are related to a variation of intensities of the diffracted electron (spots) with the energy of the electron beam. The process of determination of the structure by this technique works as `proof and error' and consists of three main parts: the measurement of the intensity spectra, the calculations for various models of atomic positions and the search for the best- t structure which is determined by an R-factor.

The first step consists of obtaining the experimental LEED pattern and all the electron beam intensities for every spot of the reciprocal lattice in the pattern. Theoretical LEED IV curves are calculated for a large number of geometrical models and these are compared with the experimental curves. The agreement is quantified by means of a reliability factor or R factor. The closest this value to zero is, the more perfect the agreement between experimental and theoretical curves. In this way, the level of precision of the crystalline structure will depend on the smallest R factor that can be achieved.

Pure metals with pure surfaces allow R factor values of around 0.1. When moving to more complex structures, these values increase. The main reason for this gradually worse agreement between theoretical and experimental LEED-IV curves lies in the approximations in conventional LEED theory, which treats the atoms as perfect spheres with constant scattering potential in between. This description results in inaccurate scattering potential for more open surfaces and organic molecules. In consequence, a precision of 1 - 2 pm can be achieved for atoms in metal surfaces, whereas the positions of atoms within organic molecules are typically determined within ±20 pm. The values of the R-factor are usually between 0.2 and 0.5, where 0.2 represents a good agreement, 0.35 a mediocre agreement and 0.5 a poor agreement.

Figure 1.13 shows an example of a typical LEED IV curve for Ir(100), which has a quasi-hexagonal unit cell. One can observe the parameters used to calculate the theoretical LEED IV curve and the best-fitted curve obtained experimentally, which has an R factor value of 0.144. The model used is also shown.

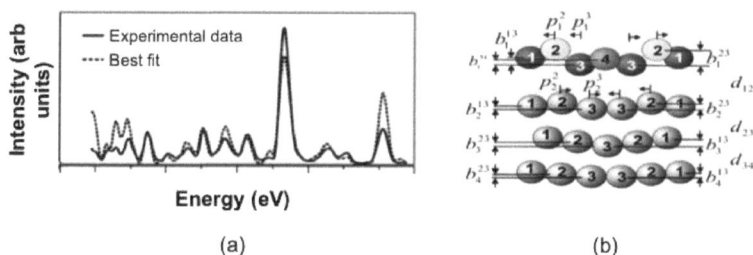

(a)   (b)

Figure 2.13: Experimental and theoretical LEED-IV curve for Ir(100) (a), and the structural parameters using for the LEED-IV theoretical curve (b). Adapted from K. Heinz and L. Hammer, Combined application of LEED and STM in surface crystallography. *J. Phys. Chem. B*, 2004, 108, 14579. Copyright: American Chemical Society (2004).

Bibliography

C. Davisson and L. Germer, The scattering of electrons by a single crystal of nickel. *Nature*, 1927, **119**, 558.

C. Davisson and L. Germer, Diffraction of electrons by a crystal of nickel. *Phys. Rev.*, 1927, **30**, 705.

R. Gehrenbeck, Electron diffraction: fifty years ago. *Phys. Today*, 1978, **31**, 34.

M. Gulde, S. Schweda, G. Storeck, M. Maiti, H. Yu, A. Wodtke, S. Schäfer, and C. Ropers, Ultrafast low-energy electron diffraction in transmission resolves polymer/graphene superstructure dynamics. *Science*, 2014, **345**, 200.

K. Heinz and L. Hammer, Combined application of LEED and STM in surface crystallography. *J. Phys. Chem. B*, 2004, **108**, 14579.

G. Held, Low-energy electron diffraction crystallography of surfaces and interfaces. *Bunsen-Magazin*, 2010, **12**, 124.

F. Jona, J. Strozier, W. Yang, Low-energy electron diffraction for surface structure analysis. *Rep. Prog. Phys.*, 1982, **45**, 527.

E. Kaufmann, *Characterization of Materials, Volume 3*, 2nd edn., Wiley Library, New Jersey (2012).

J. Manners, *Quantum Physics: An Introduction*, Institute of Physics Publishing, London (2000).

L. Meng, Y. Wang, L. Zhang, S. Du, R. Wu, L. Li, Y. Zhang, G. Li, H. Zhou, W. Hofer, and H. Gao, Buckled silicene formation on Ir(111). *Nano Lett.*, 2013, **13**, 685.

H. S. Nalwa, *Handbook of Thin Film Materials: Characterization and Spectroscopy of Thin Films, Volume 2*, Elsevier, Michigan (2002).

E. Nibbering, Low-energy electron diffraction at ultrafast speeds. *Science*, 2014, **345**, 137.

Z. Robinson, E. Ong, T. Mowll, P. Tyagi, D. Gaskill, H. Geisler, C. Ventrice, Influence of chemisorbed oxygen on the growth of graphene on Cu(100) by chemical vapor deposition. *J. Phys. Chem. C*, 2013, **117**, 23919.

F. Sedona, G. Rizzi, S. Agnoli, F. Llabrés, A. Papageorgiou, D. Ostermann, M. Sambi, P. Finetti, K. Schierbaum, and G. Granozzi, Ultrathin TiO_x films on Pt(111): a LEED, XPS, and STM investigation. *J. Phys. Chem. B*, 2005, **109**, 24411.

F. Sojka, M. Meissner, Ch. Zwick, R. Forker, and T. Fritz, Determination and correction of distortions and systematic errors in low-energy electron diffraction. *Rev. Sci. Instrum.*, 2013, **84**, 015111.

N. Spencer and J. Moore, *Encyclopedia of Chemical Physics and Physical Chemistry: Fundamentals, Volume 2*, 1st edn., Institute of Physics Publishing, London (2001).

K. Wandelt, *Surface and Interface Science, Volume 1*, Wiley-VCH, Weinheim (2012).

Chapter 3: Neutron Diffraction

Gabriela Escalera and Andrew R. Barron

Introduction

The first neutron diffraction experiment was in 1945 by Ernest O. Wollan (Figure 3.1) using the Graphite Reactor at Oak Ridge. Along with Clifford Shull (Figure 3.1) the principles of the technique were outlined. However, the concept that neutrons would diffract like X-rays was first proposed by Dana Mitchell and Philip Powers. They proposed that neutrons have a wave like structure, which is explained by the de Broglie equation,

$$\lambda = h/mv$$

where λ is the wavelength of the source usually measured in Å, h is Planck's constant, v is the velocity of the neutron, and finally m represents the mass of the neutron.

Figure 3.1: American physicists Ernest Wollan (1902 - 1984) and (standing) Clifford Shull (1915 2001).

The great majority of materials that are studied by diffraction methods are composed of crystals. X-rays where the first type of source tested with crystals in order to determine their structural characteristics. Crystals are said to be perfect structures although some of them show defects on their structure. Crystals are composed of atoms, ions or molecules, which are arranged, in a uniform repeating pattern. The basic concept to understand about crystals is that they are composed of an array of points, which are called lattice points, and the motif, which represents the body part of the crystal. Crystals are composed of a series of unit cells. A unit cell is the repeating portion of the crystal. Usually there are another eight unit cells surrounding each unit cell. Unit cells can be categorized as primitive, which have only one lattice point. This means that the unit cell will only have lattice points on the corners of the cell. This point is going to be shared with eight other unit cells. Whereas in a non-primitive cell there will also be point in the corners of the cell but in addition there will be lattice points in the faces or the interior of the cell, which similarly will be shared by other cells. The only primitive cell known is the simple crystal system and for nonprimitive cells there are known face-centered cubic, base centered cubic and body centered cubic.

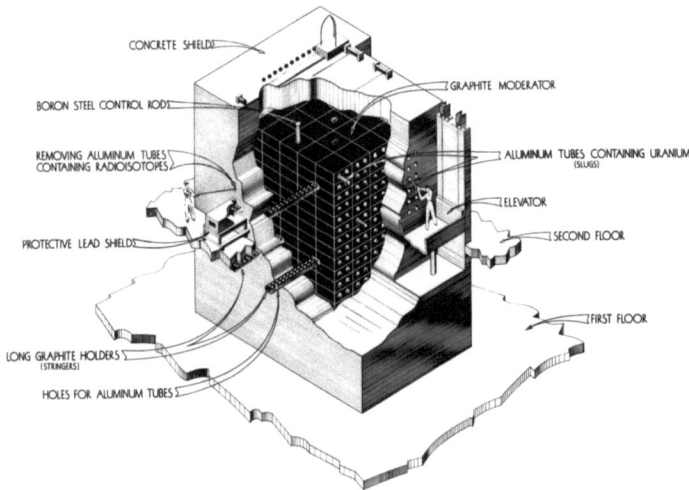

Figure 3.2: A schematic of the nuclear reactor (uranium pile) of the X-10 Graphite Reactor at Oak Ridge, Tennessee.

Crystals can be categorized depending on the arrangement of lattice points; this will generate different types of shapes. There are known seven crystal systems, which are cubic, tetragonal, orthorhombic, rhombohedral, hexagonal, monoclinic and triclinic. All of these have different angles and the axes are equally the same or different in others. Each of these types of system has different Bravais lattice.

Braggs law

Braggs Law was first derived by physicist Sir W. H. Bragg (Figure 3.3) and his son W. L Bragg (Figure 3.4) in 1913.

Figure 3.3: British physicist, chemist, mathematician and active sportsman Sir William H. Bragg (1862 - 1942).

Figure 3.4: Australian-born British physicist Sir William Lawrence Bragg (1890 - 1971).

It has been used to determine the spacing of planes and angles formed between these planes and the incident beam that had been applied to the crystal examined. Intense scattered X-rays are produced when X-rays with a set wavelength are executed to a crystal. These scattered X-rays will interfere constructively due the equality in the differences between the travel path and the integral number of the wavelength. Since crystals have repeating units, diffraction can be seen in terms of reflection from the planes of the crystals.

The incident beam, the diffracted beam and normal plane to diffraction need to lie in the same geometric plane. Figure 3.5 shows a schematic representation of how the incident beam hits the plane of the crystal and is reflected at the same angle 2θ, which the incident beam hits. Bragg's Law is mathematically expressed,

$$n\lambda = 2d \sin\theta$$

where, n = integer order of reflection, λ = wavelength, d = plane spacing.

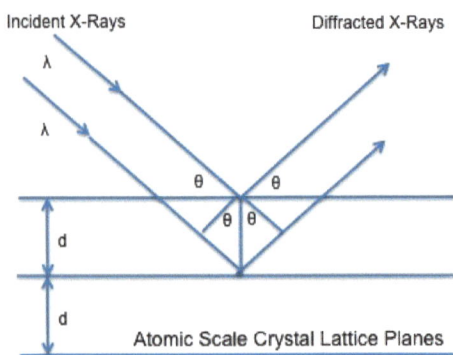

Figure 3.5: Bragg's Law construction.

Bragg's Law is essential in determining the structure of an unknown crystal. Usually the wavelength is known, and the angle of the incident beam can be measured. Having these two known values, the plane spacing of the layer of atoms or ions can be obtained. All reflections collected can be used to determine the structure of the unknown crystal material.

Bragg's Law applies similarly to neutron diffraction. The same relationship is used the only difference being is that instead of using X-rays as the source, neutrons that are ejected and hit the crystal are being examined.

Neutron diffraction

Neutrons have been studied for the determination of crystalline structures. The study of materials by neutron radiation has many advantages against the normally used such as X-rays and electrons. Neutrons are scattered by the nucleus of the atoms rather than X-rays, which are scattered by the electrons of the atoms. These generates several differences between them such as that scattering of X-rays highly depend on the atomic number of the atoms whereas neutrons depend on the properties of the nucleus. These lead to a greater and accurately identification of the unknown sample examined if neutron source is being used. The nucleus of every atom and even from isotopes of the same element is completely different. They all have different characteristics, which make neutron diffraction a great technique for identification of materials, which have similar elemental composition. In contrast, X-rays will not give an exact solution if similar characteristics are known between materials. Since the diffraction will be similar for adjacent atoms further analysis needs to be done in order to determine the structure of the unknown. Also, if the sample contains light elements such as hydrogen, it is almost impossible to determine the exact location of each of them just by X-ray diffraction or any other technique. Neutron diffraction can tell the number of light elements and the exact position of them present in the structure.

Neutrons were first discovered by James Chadwick in 1932 (Figure 3.6) when he showed that there were uncharged particles in the radiation he was using. These particles had a similar mass of the protons but did not have the same characteristics. Chadwick followed some of the predictions of Rutherford who first worked in this unknown field. Later, Walter Elsasser (Figure 3.7) designed the first neutron diffraction in 1936 and the ones responsible for the actual constructing were Halban (Figure 3.8) and Preiswerk.

Figure 7.18: English Nobel laureate in physics James Chadwick (1891 - 1974).

Figure 3.7: German born physicist Walter Maurice Elsasser (1904 - 1991).

This was first constructed for powders, but later Mitchell and Powers developed and demonstrated the single crystal system. All experiments realized in early years were developed using radium and beryllium sources. The neutron flux from these was not sufficient for the characterization of materials. Then, years passed, and neutron reactors had to be constructed in order to increase the flux of neutrons to be able to realize a complete characterization the material being examined.

Figure 3.8: French physicist Hans Heinrich von Halban (1908 - 1964).

Figure 3.9: Swiss physicist Peter Preiswerk (1907 - 1972) was co-founder of CERN (Conseil Européen pour la Recherche Nucléaire).

Between mid and late 40s neutron sources began to appear in countries such as Canada, UK and some other of Europe. Later in 1951 Shull and Wollan (Figure 3.1) presented a paper that discussed the scattering lengths of 60 elements and isotopes, which generated a broad opening of neutron diffraction for the structural information that can be obtained from neutron diffraction.

Neutron sources

The first source of neutrons for early experiments was gathered from radium and beryllium sources. The problem with this, as already mentioned, was that the flux was not enough to perform huge experiments such as the determination of the structure of an unknown material. Nuclear reactors started to emerge in early 50s and these had a great impact in the scientific field. In the 1960s neutron reactors were constructed depending on the desired flux required for the production of neutron beams. In USA the first one constructed was the High Flux Beam Reactor (HFBR). Later, this was followed by one at Oak Ridge Laboratory (HFIR) (Figure 3.10), which also was intended for isotope production and a couple of years later the ILL was built. This last one is the most powerful so far and it was built by collaboration between Germany and France. These nuclear reactors greatly increased the flux and so far, there has not been constructed any other better reactor. It has been discussed that probably the best solution to look for greater flux is to look for other approaches for the production of neutrons such as accelerator driven sources. These could greatly increase the flux of neutrons and in addition other possible experiments could be executed. The key point in these devices is spallation, which increases the number of neutrons executed from a single proton and the energy released is minimal. Currently, there are several of these around the world, but investigations continue searching for the best approach of the ejection of neutrons.

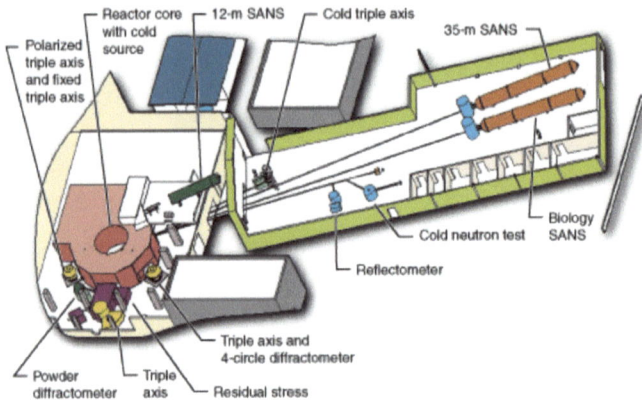

Figure 3.10: Schematic representation of HIFR. Courtesy of Oak Ridge National Laboratory, US Dept. of Energy

Neutron detectors

Although neutrons are great particles for determining complete structures of materials, they have some disadvantages. These particles experiment a reasonably weak scattering when looking especially to soft materials. This is a huge concern because there can be problems associated with the scattering of the particles which can lead to a misunderstanding in the analysis of the structure of the material.

Neutrons are particles that have the ability to penetrate through the surface of the material being examined. This is primarily due to the nuclear interaction produced from the particles and the nucleus from the material. This interaction is much greater that the one performed from the electrons, which it is only an electrostatic interaction. Also, it cannot be omitted the interaction that occurs between the electrons and the magnetic moment of the neutrons. All of these interactions discussed are of great advantage for the determination of the structure since neutrons interacts with every single nucleus in the material. The problem comes when the material is being analyzed because neutrons being uncharged materials make them difficult to detect them. For this reason, neutrons need to be reacted in order to generate charged particles, ions. Some of the reactions usually used for the detection of neutrons are:

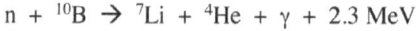

$$n + {}^3He \rightarrow {}^3H + {}^1H + 0.764 \, MeV$$

$$n + {}^{10}B \rightarrow {}^7Li + {}^4He + \gamma + 2.3 \, MeV$$

$$n + {}^6Li \rightarrow {}^4He + {}^3H + 4.79 \, MeV$$

The first two reactions apply when the detection is performed in a gas environment whereas the third one is carried out in a solid. In each of these reactions there is a large cross section, which makes them ideal for neutron capture. The neutron detection hugely depends on the velocity of the particles. As velocity increases, shorter wavelengths are produced and the less efficient the detection becomes. The particles that are executed to the material need to be as close as possible in order to have an accurate signal from the detector. These signal needs to be quickly transduced and the detector should be ready to take the next measurement.

In gas detectors the cylinder is filled up with either ^3He or BF_3. The electrons produced by the secondary ionization interact with the positively charged anode wire. One disadvantage of this detector is that it cannot be attained a desired thickness since it is very difficult to have a fixed thickness with a gas. In contrast, in scintillator detectors since detection is developed in a solid, any thickness can be obtained. The thinner the thickness of the solid the more efficient the results obtained become. Usually the absorber is ^6Li and the substrate, which detects the products, is phosphor, which exhibits luminescence. This emission of light produced from the phosphor results from the excitation of this when the ions pass thorough the scintillator. Then the signal produced is collected and transduced to an electrical signal in order to tell that a neutron has been detected.

Neutron scattering

One of the greatest features of neutron scattering is that neutrons are scattered by every single atomic nucleus in the material whereas in X-ray studies, these are scattered by the electron density. In addition, neutron can be scattered by the magnetic moment of the atoms. The intensity of the scattered neutrons will be due to the wavelength at which it is executed from the source. Figure

3.11 shows how a neutron is scattered by the target when the incident beam hits it.

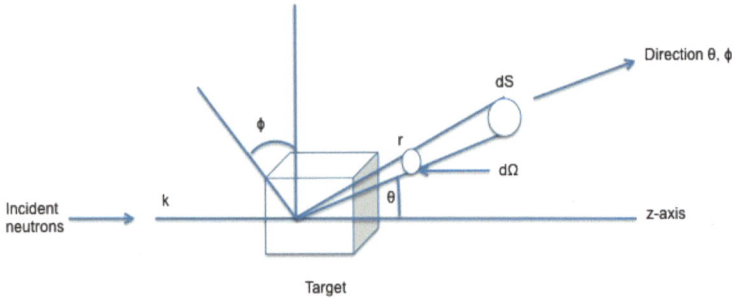

Figure 3.11: Schematic representation of scattering of neutrons when it hits the target. Adapted from W. Marshall and S. W. Lovesey, *Theory of Thermal Neutron Scattering: the use of Neutrons for the Investigation of Condensed Matter*, Clarendon Press, Oxford (1971).

The incident beam encounters the target and the scattered wave produced from the collision is detected by a detector at a defined position given by the angles θ, φ which are joined by the $d\Omega$. In this scenario there is assumed that there is no transferred energy between the nucleus of the atoms and the neutron ejected, leads to an elastic scattering.

When there is an interest in calculating the diffracted intensities, the cross-sectional area needs to be separated into scattering and absorption respectively. In relation to the energies of these there is moderately large range for constant scattering cross section. Also, there is a wide range cross sections close to the nuclear resonance. When the energies applied are less than the resonance the scattering length and scattering cross section are moved to the negative side depending on the structure being examined. This means that there is a shift on the scattering, therefore the scattering will not be in a 180 ° phase. When the energies are higher that resonance it means that the cross section will be asymptotic to the nucleus area. This will be expected for spherical structures. There is also resonance scattering when there are different isotopes because each produce different nuclear energy levels.

Coherent and incoherent scattering

Usually in every material, atoms will be arranged differently. Therefore, neutrons when scattered will be either coherently or incoherently. It is convenient to determine the differential scattering cross section, which is given by,

$$d\sigma/d\Omega \;=\; \underbrace{|\,b\,|^2 \;|\Sigma \exp(ik.r_n)\,|^2}_{\text{coherent}} \;+\; \underbrace{N\,|\,b\text{-}b\,|^2}_{\text{incoherent}}$$

where b represents the mean scattering length of the atoms, k is the scattering vector, r_n is the position of the vector of the analyzed atom and lastly N is the total number of atoms in the structure. This equation can be separated in two parts, which one corresponds to the coherent scattering and the incoherent scattering as labeled below. Usually the particles scattered will be coherent which facilitates the solution of the cross section but when there is a difference in the mean scattering length, there will be a complete arrangement of the formula and these new changes (incoherent scattering) should be considered. Incoherent scattering is usually due to the isotopes and nuclear spins of the atoms in the structure.

The ability to distinguish atoms with similar atomic number or isotopes is proportional to the square of their corresponding scattering lengths. There are already known several coherent scattering lengths of some atoms which are very similar to each other. Therefore, it makes even easier to identify by neutrons the structure of a sample. Also, neutrons can find ions of light elements because they can locate very low atomic number elements such as hydrogen. Due to the negative scattering that hydrogen develops it increases the contrast leading to a better identification of it, although it has a very large incoherent scattering which causes electrons to be removed from the incident beam applied.

Magnetic scattering

As previously mentioned, one of the greatest features about neutron diffraction is that neutrons because of their magnetic moment can interact with either the orbital or the spin magnetic moment of the material examined. Not all every single element in the periodic table can exhibit a magnetic moment. The only elements that show a magnetic moment are those, which have unpaired electrons spins. When neutrons hit the solid this produces a scattering from the magnetic moment vector as well as the scattering vector from the neutron itself. Below Figure 3.12 shows the different vectors produced when the incident beam hits the solid.

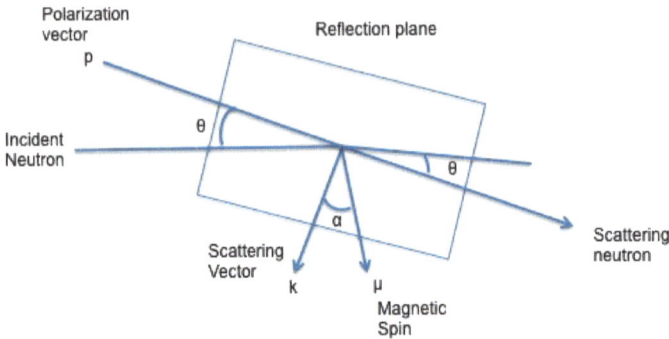

Figure 3.12: Diagram of magnetic Scattering of neutrons. Adapted from G. E. Bacon, *Neutron Diffraction*, Clarendon Press, Oxford (1975).

When looking at magnetic scattering it needs to be considered the coherent magnetic diffraction peaks where the magnetic contribution to the differential cross section is p^2q^2 for an unpolarized incident beam. Therefore, the magnetic structure amplitude will be given by,

$$F_{mag} = \Sigma p_n q_n \exp\{2\pi i(hx_n + ky_n + lz_n)\}$$

where q_n is the magnetic interaction vector, p_n is the magnetic scattering length and the rest of the terms are used to know the position of the atoms in the unit cell. When this term F_{mag} is squared, the result is the intensity of magnetic contribution from the peak analyzed. This equation only applies to those elements which have atoms that develop a magnetic moment.

Magnetic diffraction becomes very important due to its d-spacing dependence. Due to the greater effect produced from the electrons in magnetic scattering the forward scattering has a greater strength than the backward scattering. There can also be developed similar as in X-ray, interference between the atoms which makes structure factor also be considered. These interference effects could be produced by the wide range in difference between the electron distribution and the wavelength of the thermal neutrons. This factor quickly decreases as compared to X-rays because the beam only interacts with the outer electrons of the atoms.

Sample preparation and environment

In neutron diffraction there is not a unique protocol of factors that should be considered such as temperature, electric field and pressure to name a few. Depending on the type of material and data that has been looked the parameters are assigned. There can be reached very high temperatures such as 1800 K or it can go as low as 4 K. Usually to get to these extreme temperatures a special furnace capable of reaching these temperatures needs to be used. For example, one of the most common used is the He refrigerator when working with very low temperatures. For high temperatures, there are used furnaces with a heating element cylinder such as vanadium (V), niobium (Nb), tantalum (Ta) or tungsten (W) that is attached to copper bars which hold the sample. Figure 3.13 shows the design for the vacuum furnaces used for the analysis. The metal that works best at the desired temperature range will be the one chosen as the heating element. The metal that is commonly used is vanadium because it prevents the contribution of other factors such as coherent scattering. Although with this metal this type of scattering is almost completely reduced. Other important factor about this furnace is that the material been examined should not decompose under vacuum conditions. The crystal needs to be as stable as possible when it is being analyzed. When samples are not able to persist at a vacuum environment, they are heated in the presence of several gases such as nitrogen or argon.

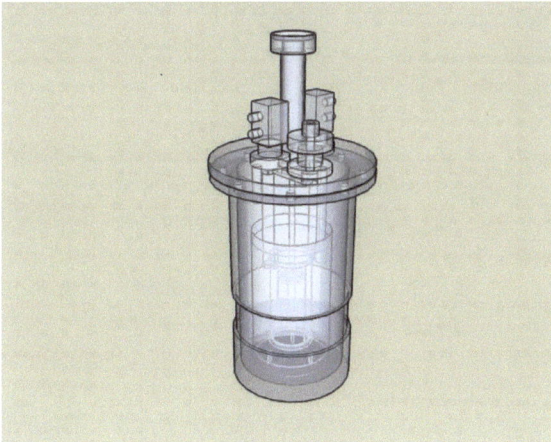

Figure 3.13: Metallic chamber which holds the sample. Courtesy of Nuclear Physics Institute.

Usually in order to prepare the samples that are being examined in neutron diffraction it is needed large crystals rather small ones as the one needed for X-ray studies. This one of the main disadvantages of this instrument. Most experiments are carried out using a four-circle diffractometer. The main reason being is because several experiments are carried out using very low temperatures and in order to achieve a good spectrum it is needed the He refrigerator. First, the crystal being analyzed is mounted on a quartz slide, which needs to be a couple millimeters in size. Then, it is inserted into the sample holder, which is chosen depending on the temperatures wanted to be reached. In addition, neutrons can also analyze powder samples and in order to prepare the sample for these they need to be completely rendered into very ne powders and then inserted into the quartz slide similarly to the crystal structures. The main concern with this method is that when samples are grounded into powders the structure of the sample being examined can be altered.

Summary

Neutron diffraction is a great technique used for complete characterization of molecules involving light elements and also very useful for the ones that have different isotopes in the structure. Due to the fact that neutrons interact with the nucleus of the atoms rather than with the outer electrons of the atoms such as X-rays, it leads to a more reliable data. In addition, due to the magnetic properties of the neutrons there can be characterized magnetic compounds due to the magnetic moment that neutrons develop. There are several disadvantages as well, one of the most critical is that there needs to be a good amount of sample in order to be analyzed by this technique. Also, great amounts of energy are needed to produce large amounts of neutrons. There are several powerful neutron sources that have been developed in order to conduct studies of largest molecules and a smaller quantity of sample. However, there is still the need of devices which can produce a great amount of flux to analyze more sophisticated samples. Neutron diffraction has been widely studied due to the fact that it works together with X-rays studies for the characterization of crystalline samples. The properties and advantages of this technique can greatly increase if some of the disadvantages are solved. For example, the study of molecules which exhibit some type of molecular force can be characterized. This will be because neutrons can precisely locate hydrogen atoms in a sample. Neutrons have given a better answer to the chemical interactions that are present in every single molecule, whereas X-rays help to give an idea of the macromolecular structure of the samples being examined.

Bibliography

G. E. Bacon, *Neutron Diffraction*, Clarendon Press, Oxford (1975).

R. K. Crawford, Position-sensitive detection of slow neutrons - survey of fundamental principles. *SPIE*, 1992, **1737**, 210.

B. D. Cullity, Elements of X-Ray Diffraction, 3rd edn., Prentice Hall, New Jersey (2001).

W. Marshall and S. W. Lovesey, *Theory of Thermal Neutron Scattering: the use of Neutrons for the Investigation of Condensed Matter*, Clarendon Press, Oxford (1971).

D. P. Mitchel and P. N. Powers, Bragg reflection of slow neutrons. *Phys. Rev.*, 1936, **50**, 486.

P. N. Powers, The magnetic scattering of neutrons. *Phys. Rev.*, 1938, **54**, 827.

C. G. Shull and E. O. Wollan, Coherent scattering amplitudes as determined by neutron diffraction. *Phys. Rev.*, 1951, **81**, 527.

Chapter 4: X-ray Absorption Fine Structure Spectroscopy

Natalia Gonzalez Pech and Andrew R. Barron

Introduction

X-ray absorption fine structure (XAFS) spectroscopy includes both X-ray absorption near edge structure (XANES) and extended X-ray absorption fine structure (EXAFS) spectroscopies. The difference between both techniques is the area to analyze, as shown Figure 4.1 and the information each technique provides. The complete XAFS spectrum is collected across an energy range of around 200 eV before the absorption edge of interest and until 1000 eV after it (Figure 4.1). The absorption edge is defined as the X-ray energy when the absorption coefficient has a pronounced increasing. This energy is equal to the energy required to excite an electron to an unoccupied orbital.

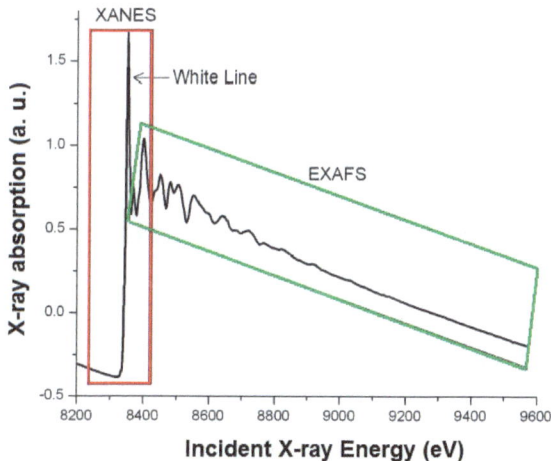

Figure 4.1: Characteristic spectra areas for X-ray absorption near edge structure (XANES) and extended X-ray absorption fine structure (EXAFS) spectroscopies. Adapted from S. D. Kelly, D. Hesterberg, and B. Ravel in *Methods of Soil Analysis: Part 5, Mineralogical Methods*, Ed. A. L. Urely and R. Drees, Soil Science Society of America Book Series, Madison (2008).

X-ray absorption near edge structure (XANES) is used to determine the valence state and coordination geometry, whereas extended X-ray absorption fine structure (EXAFS) is used to determine the local molecular structure of a particular element in a sample.

X-ray absorption near edge structure (XANES) spectra

XANES is the part of the absorption spectrum closer an absorption edge. It covers from approximately -50 eV to +200 eV relative to the edge energy (Figure 4.1). Because the shape of the absorption edge is related to the density of states available for the excitation of the photoelectron, the binding geometry and the oxidation state of the atom affect the XANES part of the absorption spectrum.

Before the absorption edge, there is a linear and smooth area. Then, the edge appears as a step, which can have other extra shapes as isolated peaks, shoulders or a white line, which is a strong peak onto the edge. Those shapes give some information about the atom. For example, the presence of a white line indicates that after the electron releasing, the atomic states of the element are confined by the potential it feels. This peak sharp would be smoothed if the atom could enter to any kind of resonance. Important information is given because of the absorption edge position. Atoms with higher oxidation state have fewer electrons than protons, so, the energy states of the remaining electrons are lowered slightly, which causes a shift of the absorption edge energy up to several eV to a higher X-ray energy.

Extended X-ray absorption fine structure (EXAFS) spectra

The EXAFS part of the spectrum is the oscillatory part of the absorption coefficient above around 1000 eV of the absorption edge. This region is used to determine the molecular bonding environments of the elements. EXAFS gives information about the types and numbers of atoms in coordination a specific atom and their inter-atomic distances. The atoms at the same radial distance from a determinate atom form a shell. The number of the atoms in the shell is the coordination number (e.g., Figure 4.2).

An EXAFS signal is given by the photoelectron scattering generated for the center atom. The phase of the signal is determinate by the distance and the path the photoelectrons travel. A simple scheme of the different paths is shown by Figure 4.3. In the case of two shells around the centered atom, there is a degeneracy of four for the path between the main atom to the first shell, a degeneracy of four for the path between the main atom to the second shell, and a degeneracy of eight for the path between the main atom to the first shell, to the second one and to the center atom.

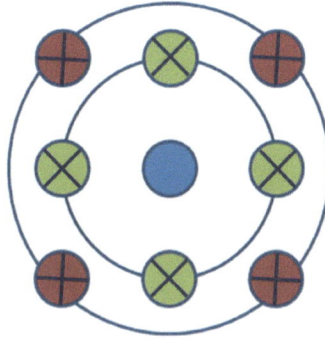

Figure 4.2: A schematic representation of coordination number in different layers in which there are two shells around the center atom. Both shells, green (x) and red (+), have coordination numbers of 4, but the radial distance of the red one (+) is bigger than the green one (x). Based on S. D. Kelly, D. Hester-berg, and B. Ravel in *Methods of Soil Analysis: Part 5, Mineralogical Methods*, Ed. A. L. Urely and R. Drees, Soil Science Society of America Book Series, Madison (2008).

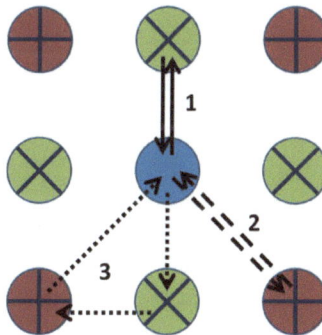

Figure 4.3: A two shell diagram in which there are three kinds of paths. From the center atom to the green one (x) and then going back (1); from the center atom to the red one (+) and the going back (2); and from the center atom to the first shell to the second one, and the returning to the center atom (3). Based on S. D. Kelly, D. Hesterberg, and B. Ravel in *Methods of Soil Analysis: Part 5, Mineralogical Methods*, Ed. A. L. Urely and R. Drees, Soil Science Society of America Book Series, Madison (2008).

The analysis of EXAFS spectra is accomplished using Fourier transformation to t the data to the EXAFS equation. The EXAFS equation is a sum of the contribution from all scattering paths of the photoelectrons,

$$\chi(k) = \sum_i \chi_i(k)$$

where each path is given by,

$$\chi_i(k) \equiv \frac{(N_i S_0^2) F_{eff,i}(k)}{k R_i^2} \sin[2kR_i + \phi_i(k)] e^{-2\sigma^2 k^2} e^{-2R_i/\lambda(k)}$$

The terms $F_{eff,i}(k)$, $\phi_i(k)$, and $\lambda_i(k)$ are the effective scattering amplitude of the photoelectron, the phase shift of the photoelectron, and the mean free path of the photoelectron, respectively. The term R_i is the half path length of the photoelectron (the distance between the centered atom and a coordinating atom for a single-scattering event). And the k^2 is given by the,

$$k_2 = [2m_e(E - E_0 + \Delta E_0)]/\hbar$$

The remaining variables are frequently determined by modeling the EXAFS spectrum.

XAFS analysis for arsenic adsorption onto iron oxides

The absorption of arsenic species onto iron oxide offers an example of the information that can be obtained by EXAFS. Because the huge impact that the presence of arsenic in water can produce in societies there is a lot of re-search in the adsorption of arsenic in several kinds of materials, in particular nano materials. Some of the materials more promising for this kind of appli-cations are iron oxides. The elucidation of the mechanism of arsenic coordination onto the surfaces of those materials has been studied lately using X-ray absorption spectroscopy.

There are several ways how arsenate (AsO_4^{3-}, Figure 4.4) can be adsorbed onto the surfaces. Figure 4.5 shows the three ways that Sherman proposes arsenate can be adsorbed onto goethite (α-FeOOH): bidentate corner sharing (2C), bidentate edge sharing (2E) and monodentate corner-sharing (1V) shapes. Figure 4.6 shows that the bidentate corner sharing (2C) is the configuration that corresponds with the calculated parameters not only for goethite, but for several iron oxides.

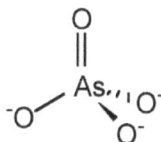

Figure 4.4: Structure of the arsenate anion.

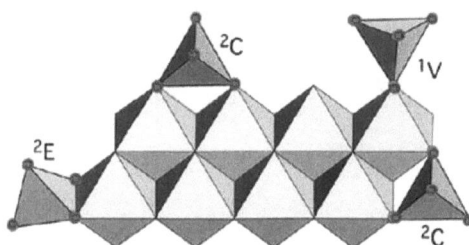

Figure 4.5: Possible configurations of arsenate onto goethite. The tetrahedral with the small spheres represents the arsenate ions. Adapted from D. M. Sherman and S. R. Randal, Surface complexation of arsenic(V) to iron(III) (hydr)oxides: structural mechanism from *ab initio* molecular geometries and EXAFS spectroscopy. *Geochim. Cosmochim. Ac.* 2003, 67, 4223. Copyright: Elsevier (2003).

Several studies have confirmed that the bidentate corner sharing (2C) is the one present in the arsenate adsorption but also one similar, a tridentate corner sharing complex (3C), for the arsenite adsorption onto most of iron oxides as shows Figure 4.7. Table 4.1 shows the coordination numbers and distances reported in the literature for the As(III) and As(V) onto goethite.

As	CN As-O	RAs-O (Å)	CN As-Fe	RAs-Fe (Å)
III	3.06 ±0.03	1.79 ±0.8	2.57 ±0.01	3.34 ±3
	3.19	1.77 ±1	1.4	3.34 ±5
	3	1.78	2	3.55 ±5
V	1.03.0	1.63 - 1.70	2	3.30
	4.6	1.68		3.55 ±5

Table 4.1: Coordination numbers (CN) and inter-atomic distances (R) reported in the literature for the As(III) and As(V) adsorption onto goethite.

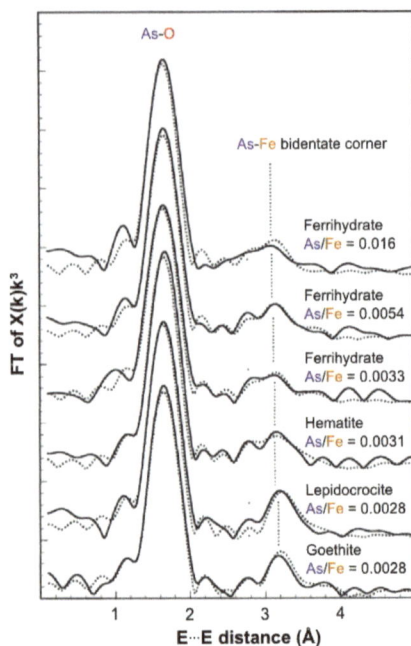

Figure 4.6: Fourier transforms of the EXAFS for arsenate adsorbed onto goethite, lepidocrocite, hematite and ferrihydrite. Adapted from D. M. Sherman and S. R. Randal, Surface complexation of arsenic(V) to iron(III) (hydr)oxides: structural mechanism from ab initio molecular geometries and EXAFS spectroscopy. *Geochim. Cosmochim. Ac.* 2003, 67, 4223. Copyright: Elsevier (2003).

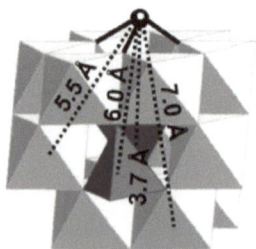

Figure 4.7: Proposed structural model for arsenic(III) tridante. Adapted from G. Morin, Y. Wang, G. Ona-Nguema, F. Juillot, G. Calas, N. Menguy, E. Aubry, J. R. Bargar, and G. E. Brown, EXAFS and HRTEM evidence for As(III)-containing surface precipitates on nanocrystalline magnetite: implications for As sequestration. *Langmuir*, 2009, 25, 9119. Copyright: American Chemical Society (2009).

Bibliography

M. Au an, J. Rose, O. Proux, D. Borschneck, A. Masion, P. Chaurand, J. L. Hazemann, C. Chaneac, J. P. Jolivet, M. R. Wiesner, A. Van Geen, and J. Y. Bottero, *Langmuir*, 2008, **24**, 3215.

G. Bunker. Introduction to XAFS: A practical guide to X-ray Absorption Fine Structure Spectroscopy, Cambridge University Press, Cambridge (2010).

B. A. Manning, and S. E. Fendorf, and S.Goldberg, Surface structures and stability of arsenic(III) on goethite: spectroscopic evidence for inner-sphere complexes. *Environ. Sci. Technol.*, 1998, **32**, 2383.

S. D. Kelly, D. Hesterberg, and B. Ravel in *Methods of Soil Analysis: Part 5, Mineralogical Methods*, Ed. A. L. Urely and R. Drees, Soil Science Society of America Book Series, Madison (2008).

G. Morin, Y. Wang, G. Ona-Nguema, F. Juillot, G. Calas, N. Menguy, E. Aubry, J. R. Bargar, and G. E. Brown, EXAFS and HRTEM evidence for As(III)-containing surface precipitates on nanocrystalline magnetite: implications for As sequestration. *Langmuir*, 2009, **25**, 9119.

G. Ona-Nguena, G. Morin F. Juillot, G. Calas, and G. E. Brown, EXAFS analysis of arsenite adsorption onto two-line ferrihydrite, hematite, goethite, and lepidocrocite. *Environ. Sci. Technol.*, 2005, **39**, 9147.

G. Ona-Nguena, G. Morin, Y. Wang, N. Menguy, F. Juillot, L, Olivi, G. Aquilanti, M. Abdelmoula, C. Ruby, J. R. Bargar, F. Guyot, G. Calas, and G. E. Brown, Jr., Arsenite sequestration at the surface of nano-Fe(OH)$_2$, ferrous-carbonate hydroxide, and green-rust after bioreduction of arsenic-sorbed lepidocrocite by *Shewanella putrefaciens*. *Geochim. Cosmochim. Ac.*, 2009, **73**, 1359.

J. Rose, M. M. Cortalezzi-Fidalgo, S. Moustier, C. Magnetto, C. D. Jones, A. R. Barron, M. R. Wiesner, and J.-Y. Bottero, Synthesis and characterization of carboxylate–FeOOH nanoparticles (ferroxanes) and ferroxane-derived ceramics. *Chem. Mater.*, 2002, **14**, 621.

D. M. Sherman and S. R. Randal, Surface complexation of arsenic(V) to iron(III) (hydr)oxides: structural mechanism from *ab initio* molecular geometries and EXAFS spectroscopy. *Geochim. Cosmochim. Ac.* 2003, **67**, 4223.

M. Stachowicz, T. Hiemstra, and W. H. Van Riemsdijk, Surface speciation of As(III) and As(V) in relation to charge distribution. *J. Colloid. Interface. Sci.*, 2005, **302**, 62.

G. A. Waychunas, B. A. Rea, C. C. Fuller, and J. A. Davis, Surface chemistry of ferrihydrite: Part 1. EXAFS studies of the geometry of coprecipitated and adsorbed arsenate. *Geochim. Cosmochim. Ac.*, 1993, **57**, 2251.

Chapter 5: Determination of Ligand Steric Bulk

Kexin Ling, Pavan M. V. Raja and Andrew R. Barron

Introduction

Steric effects are a kind of non-bonding interactions, which may influence the shape and reactivity of ions and molecules. The origin of steric effects is mostly repulsion from overlapped electron cloud. Coordination compounds consist of a center atom or ion (usually metallic) and a surrounding array of bound molecules or ions, which are called ligands. Ligands can vary from small, inorganic molecules or ions such as Cl^-, H_2O, and CO, also large organic molecules like branched alkanes, phosphines, or conjugated structures. Those big organic ligands are often steric hindered, and these steric effects do affect physical and chemical properties of these complexes a lot. In this module, we are going to talk about the synthesis methods, characterization techniques, and special properties and applications of coordination complexes with ligand steric bulk.

Synthesis methods

Methods using for synthesis these compounds depend on the physical and chemical properties of the target compounds. For example, many coordination compounds are very sensitive to water, so the synthesis process is often carried out using Schlenk techniques under inert gas atmosphere and using anhydrous solvents. Most often used methods include solid state synthesis, solution-based synthesis and solvothermal synthesis. Many reactions are based on ligand substitutions, which can be presented as:

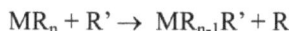

$$MR_n + R' \rightarrow MR_{n-1}R' + R$$

There also reactions to form adducts for electron deficient metal centers like Group 13 metal compounds:

$$MR_n + R' \rightarrow MR_nR'$$

Effects on coordination compounds formation

Steric repulsion between ligands significantly influence the formation of co-ordination complexes. Generally, the steric bulk ligands will slow down ligand substitution process at the same metal center, as well as prevent the aggregations into complexes with high coordination metal centers. This kind of affect also related to the kind and size of center metal.

The synthesis of tetrameric Al(I) cyclopentadienyl compound, Al_4Cp*_4 ($Cp*$ = C_5Me_5) and new $[AlR]_4$ (R = C_5Me_4Pr, $C_5Me_4Pr_i$) tetramers. By ^{27}Al NMR data and density functional theory (DFT) calculation of thermochemistry properties, the new compounds showed that they are enthalpically more stable in the form of tetramers than Al_4Cp*_4 as a result of non-covalent interactions between the bulk ligand groups. Figure 5.1 shows the structures, while Table 5.1 gives the calculated and experimental enthalpy changes (ΔH_{tet}).

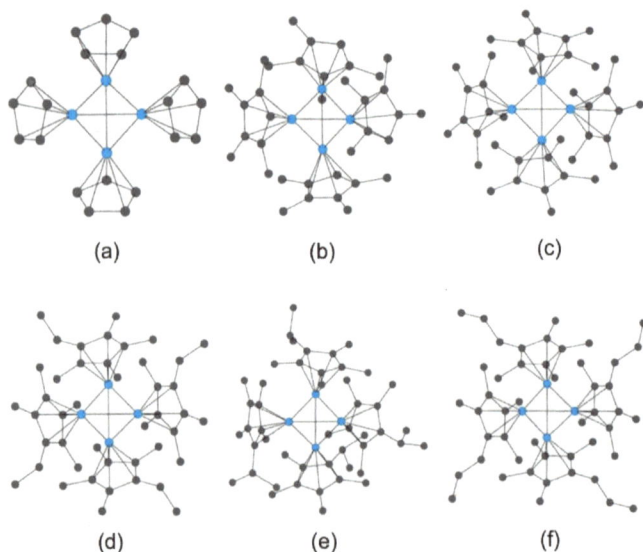

Figure 5.1: Six substituted cyclopentadienyl [AlR]₄ compounds studied in the order of increasing steric bulk: (a) $[Al(Cp)]_4$ (Cp = C_5H_5), (b)) $[Al(Cp')]_4$ (Cp' = C_5HMe_4), (c)) $[Al(Cp*)]_4$ ($Cp*$ = C_5Me_5), (d) $[Al(Cp*Et)]_4$ ($Cp*Et$ = C_5Me_4Et), (e)) $[Al(Cp*^iPr)]_4$ ($Cp*^iPr$ = $C_5Me_4^iPr$), and (f)) $[Al(Cp*^nPr)]_4$ ($Cp*^nPr$ = $C_5Me_4^nPr$). Adapted from W. W. Tomlinson, D. H. Mayo, R. M. Wilson, and J. P. Hooper, The role of ligand steric bulk in new monovalent aluminum compounds. *J. Phys. Chem. A*, 2017, 121, 4678. Copyright: American Chemical Society (2017).

Ligand	Calculated ΔH_{tet} (kJ/mol)	Experimental ΔH_{tet} (kJ/mol)
C_5H_5 (Cp)	-106.7	-
C_5HMe_4 (Cp')	-153.1	-
C_5Me_5 (Cp*)	-148.4	-150
C_5Me_4Et (Cp*Et)	-151.5	-
$C_5Me_4{}^iPr$ (Cp*^{i}Pr)	-155.8	-158
$C_5Me_4{}^nPr$ (Cp*^{n}Pr)	-158.7	-160

Table 5.1: The calculated and experimental (ΔH_{tet}) for substituted cyclopentadienyl [AlR]₄ compounds, whose structures are shown in Figure 5.1. Data from W. W. Tomlinson, D. H. Mayo, R. M. Wilson, and J. P. Hooper, The role of ligand steric bulk in new monovalent aluminum compounds. *J. Phys. Chem. A*, 2017, 121, 4678.

Characterization method

Cone angle

Ligand steric effect can be measured by measuring Tolman cone angle (θ). Tolman cone angle, also called ligand cone angle, is defined as the solid angle, which is formed by the metal at the vertex and the atoms at the perimeter of cone (Figure 5.2a). This concept was first introduced by Chadwick A. Tolman (Figure 5.3), which was originally applied to phosphines, but this method can be used to measure any ligand, especially ligand that are steric bulky.

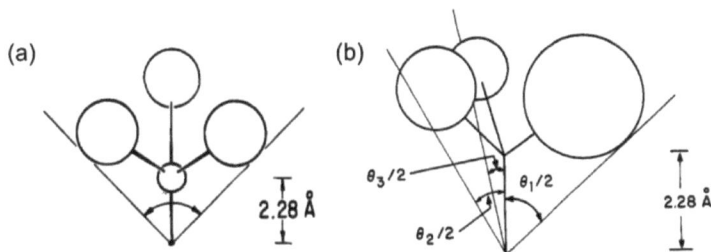

Figure 5.2: The cone angle (a) and method of measuring cone angles for asymmetrical ligands (b). Adapted from C. A. Tolman, Steric effects of phosphorus ligands in organometallic chemistry and homogeneous catalysis. *Chem. Rev.*, 1977, 77, 313. Copyright: American Chemical Society (1977).

Figure 5.3: American chemist Chadwick A. Tolman (1938-).

Ligand cone angle can be measured from accurate physical structure modules by measuring bond lengths and bond angles and do calculation. Symmetric ligands, like PR_3, can be easily visualized. For asymmetric ligands, like $PR_1R_2R_3$, the cone angle can be measured by using a module to minimize the sum of half angles (Figure 5.2b), which is shown in,

$$\theta = \frac{2}{3} \sum_{i=1}^{3} \frac{\theta_i}{2}$$

Besides the pure data of ligand cone angle, scientists are more interested in how ligands with different cone angles will affect chemical and physical properties of coordination compounds. More techniques are used to study ligand steric bulk.

Single-crystal X-ray crystallography

Using single-crystal X-ray crystallography, crystals are studied under an X-ray beam at different angles. By collecting these diffraction signals and processed by Fourier transform, information about atom positions, bond lengths and angles, and relative orientations of molecules can all be acquired. Coordination com- pounds contain metal-ligand bonds, and are relatively easy to crystallize, which makes single-crystal X-ray crystallography a perfect way to determine the structure and orientation of ligands.

A consideration of the series of aluminum compounds, $[R_2Al(\mu\text{-}OCH_2CH_2OMe)_2]_2$, where R = Me, Et, ^{i}Bu and ^{t}Bu. (Figure 5.4a - c). By studying crystallography data, the only difference between these three compounds is the extent of $Al\cdots O_{(ether)}$ interaction. This distance in $[Me_2Al(\mu\text{-}OCH_2CH_2OMe)_2]_2$, and $[^iBu_2Al(\mu\text{-}OCH_2CH_2OMe)_2]_2$, is longer than the ordinary dative Lewis acid-base interactions, which is between 1.90 and 2.20 Å. In $[^tBu_2Al(\mu\text{-}OCH_2CH_2OMe)_2]_2$, this distance is even longer (2.74 Å), which is close to the limit of van der Waal interaction. Figure 5.4d shows this interaction is dependent on the steric bulk of aluminum alkyl, as the cone angles are tert-butyl ($\theta = 126°$), iso-butyl ($\theta = 108°$), and methyl ($\theta = 90°$).

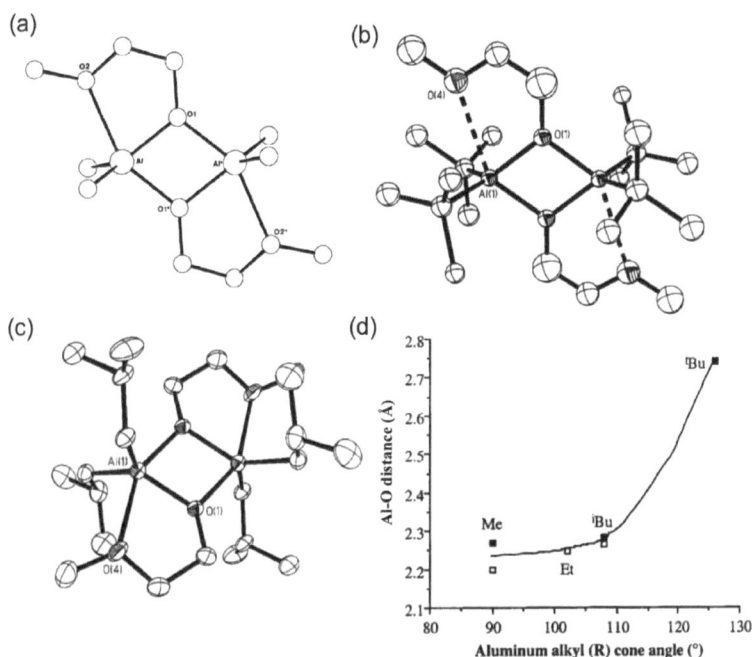

Figure 5.4: Molecular structures of (a) $[Me_2Al(\mu\text{-}OCH_2CH_2OMe)_2]$, (b) $[(^tBu)_2Al(\mu\text{-}OCH_2CH_2OMe)_2]$, and (c) $[(^iBu)_2Al(\mu\text{-}OCH_2CH_2OMe)_2]$. Hydrogen atoms are omitted. (d) Plot of $Al\cdots O_{(ether)}$ distance in $[R_2Al(\mu\text{-}OCH_2CH_2OMe)_2]$ as a function of the aluminum alkyl (R) cone angle. (a) Adapted from R. Benn, A. Rufińska, H. Lehmkuhl, E. Janssen, and C. Krüger, ^{27}Al-NMR spectroscopy: a probe for three-, four-, five-, and sixfold coordinated al atoms in organoaluminum compounds. *Angew. Chem. Int. Ed.*, 1983, 22, 779. Copyright (2003) John Wiley and Sons. (b-d) Adapted with permission from J. A. Francis, C. N. McMahon, S. G. Bott, and A. R. Barron, Steric effects in aluminum compounds containing monoanionic potentially bidentate ligands: toward a quantitative measure of steric bulk. *Organometallics*, 1999, 18, 4399. Copyright (1999) American Chemical Society.

Nuclear magnetic resonance spectroscopy

Nuclear magnetic resonance (NMR) spectroscopy is a widely used method for organic chemists. It is the most reliable and convenient way to determine the structure of a compound. NMR works based on the intrinsic spin of nucleus in external magnetic field. From chemical shift and coupling on NMR spectrum, chemical environment information of each detected atom can be acquired. Detailed information about basic NMR theory, instrument and operation, and NMR spectrum of different nuclides can refer to NMR properties of the elements.

Adducts of trimethylaluminum with phosphine ligands can be used as a probe of how the chemical shifts changes both in ^{31}P NMR and ^{13}C NMR related to the ligand steric bulk. Figure 7.38a shows a linear relationship between ^{31}P coordinate chemical shift change ($\Delta\delta$) and phosphine cone angle (θ). For those ligands with a cone angle $\theta < 142°$, $\Delta\delta$ is negative, so R-P-R angle should be opened compared with free phosphine. The other ligands with $\theta > 142°$ should have a more accurate R-P-R angle due to the same reason. This is probably due to the steric repulsion of aluminum methyl groups. The author then plotted aluminum methyl ^{13}C NMR shift (δ) as a function of phosphine cone angle (θ). A similar linear trend was also found (Figure 5.5b). Although electronic effect should also be considered as a factor to influence the chemical shift, the increasing phosphine steric bulk was convinced to be the main reason that both cause the decrease in $\Delta(^{31}P)$ and down field in ^{13}C shift.

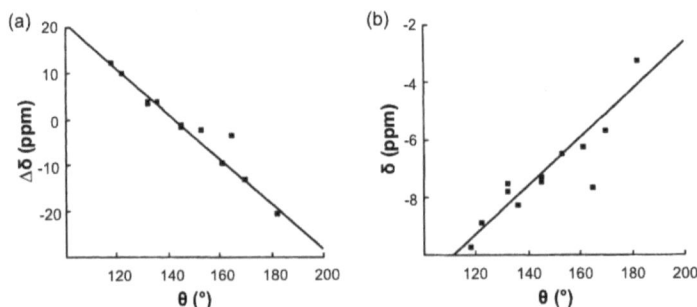

Figure 5.5: (a) Plot of ^{31}P chemical shift change ($\Delta\delta$) as a function of phosphine cone angle (θ): $\Delta = 69.84(\theta)\ 0.490$, $R^2 = 0.99$, (b) plot of aluminum methyl ^{13}C NMR shift (δ) as a function of phosphine cone angle (θ): $\delta = -19.29(\theta)\ 0.0840$ ($R^2 = 0.96$). Adapted from A. R. Barron, Adducts of trimethylaluminium with phosphine ligands; electronic and steric effects. *J. Chem. Soc., Dalton Trans.*, 1988, 12, 3047. Copyright: Royal Society of Chemistry (1988).

NMR and X-ray crystallography are complementary at some time when analyzing structures and dynamics. Both of them can give information of average properties of the specimen. However, solution NMR data represent an average of randomly oriented molecules in solution, while X-rag crystallography data gives average molecular orientation information in solid, periodic crystal lattice. This difference is particularly significant when the two measurements are applied to study ligand steric effects, which should get very detailed, accurate structure information. These two techniques do produce information from different perspectives and are both important for learning structures and dynamics of chemical compounds.

Applications

Coordination compounds are known to be useful in catalysis when they are coordinated with an electron donor. However, catalytic activity of complexes with ligands that have similar electron donor abilities may vary a lot. Steric effects are considered to be involved in improving catalytic performance. Pt complexes are a kind of powerful catalysis used for carbon-carbon bond forming. Figure 5.6 shows structures of a series of Pd-PEPPSI-R compounds.

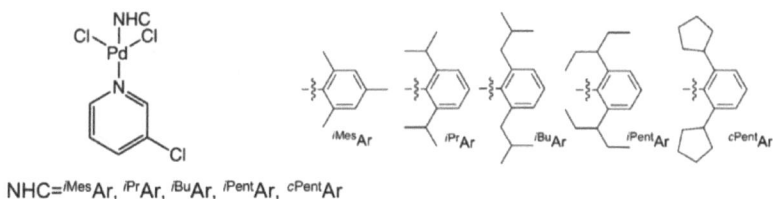

NHC=^{iMes}Ar, ^{iPr}Ar, ^{iBu}Ar, ^{iPent}Ar, ^{cPent}Ar

Figure 5.6: A series of monoligated palladium-N-heterocyclic carbene (NHC) complexes used in carbon-carbon bond formation reactions. Adapted with permission from M. G. Organ, S. Calimsiz, M. Sayah, K. H. Hoi, and A. J. Lough, Pd-PEPPSI-iPent: an active, sterically demanding cross-coupling catalyst and its application in the synthesis of tetra-ortho-substituted biaryls. *Angew. Chem. Int. Ed.*, 2009, 48, 2383. Copyright: John Wiley and Sons (2009).

An increase of steric bulk from Pd-PEPPSI-^{iMes}Ar to Pd-PEPPSI-^{cPent}Ar can be observed. Suzuki-Miyuara coupling between 2-methoxynapthalene bromide and 2,6- dimethylphenyl boronic acid was used to explore the effect of catalyst bulk,

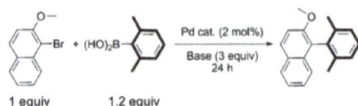

By screening base/solvent systems, Pd-PEPPSI-^{iPr}Ar and Pd-PEPPSI-^{iPent}Ar showed better performance under many base/solvent systems and were selected for further tests. From the experimental data, Pd-PEPPSI-^{iPent}Ar catalyzed coupling of both aryl bromides and aryl chlorides have a higher yield than Pd-PEPPSI-^{iPent}Ar catalyzed coupling under the same reaction condition. Figure 5.7 is the effect of temperature and reaction time on module Suzuki-Miyaura reaction using Pd-PEPPSI-^{iP}Ar and Pd-PEPPSI-^{iPent}Ar. Which shows Pd-PEPPSI-^{iPent}Ar has a better performance at both higher and lower temperature with a similar reaction speed. However, steric bulk is not the only factor that effect catalytic performance. By comparing Pd-PEPPSI-^{iPent}Ar and Pd-PEPPSI-^{cPent}Ar, it shows that conformational flexibility in the alkyl substituent is an essential while increasing steric bulk.

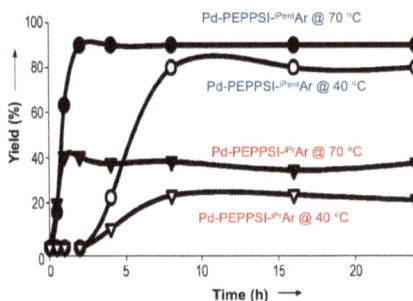

Figure 5.7: Effect of the temperature and reaction time on the model Suzuki Miyaura reaction utilizing Pd-PEPPSI complexes. Adapted with permission from M. G. Organ, S. Calimsiz, M. Sayah, K. H. Hoi, and A. J. Lough, Pd-PEPPSI-iPent: an active, sterically demanding cross-coupling catalyst and its application in the synthesis of tetra-ortho-substituted biaryls. *Angew. Chem. Int. Ed.*, 2009, 48, 2383. Copyright: John Wiley and Sons (2009).

Bibliography

J. J. Allen, C. E. Hamilton, and A. R. Barron, Synthesis and characterization of aryl-substituted *bis*(2-pyridyl)amines and their copper olefin complexes: investigation of remote steric control over olefin binding. *Dalton Trans.*, 2010, 11451.

J. J. Allen and A. R. Barron, Demonstration of remote steric differentiation of *cis/trans* alkene coordination in copper(I) complexes of aryl substituted *bis*(2-pyridyl)amine. *Dalton Trans.*, 2011, **40**, 1189.

A. R. Barron, Adducts of trimethylaluminium with phosphine ligands; electronic and steric effects. *J. Chem. Soc., Dalton Trans.*, 1988, **12**, 3047.

R. Benn, A. Rufińska, H. Lehmkuhl, E. Janssen, and C. Krüger, ^{27}Al-NMR spectroscopy: a probe for three-, four-, five-, and sixfold coordinated al atoms in organoaluminum compounds. *Angew. Chem. Int. Ed.*, 1983, **22**, 779.

D. Dange, J. Li, C. Schenk, H. Schnöckel, and C. Jones, Monomeric group 13 metal(I) amides: enforcing one-coordination through extreme ligand steric bulk. *Inorg. Chem.*, 2012, **51**, 13050.

J. A. Francis, C. N. McMahon, S. G. Bott, and A. R. Barron, Steric effects in aluminum compounds containing monoanionic potentially bidentate ligands: toward a quantitative measure of steric bulk. *Organometallics*, 1999, **18**, 4399.

Y. Koide, S. G. Bott, and A. R. Barron, Reaction of amines with [('Bu)Al(μ_3-O)]$_6$: determination of the steric limitation of a latent Lewis acid. *Organometallics*, 1996, **15**, 5514.

C. N. McMahon, S. G. Bott, and A. R. Barron, Steric effects in aluminum compounds containing monoanionic potentially bidentate ligands: effect of the steric bulk at the α-carbon. *Main Group Chem.*, 1999, **3**, 43.

M. G. Organ, S. Calimsiz, M. Sayah, K. H. Hoi, and A. J. Lough, Pd-PEPPSI-iPent: an active, sterically demanding cross-coupling catalyst and its application in the synthesis of tetra-ortho-substituted biaryls. *Angew. Chem. Int. Ed.*, 2009, **48**, 2383.

A. Tolman, Phosphorus ligand exchange equilibriums on zerovalent nickel. Dominant role for steric effects. *J. Am. Chem. Soc.*, 1970, **92**, 2956.

C. A. Tolman, Steric effects of phosphorus ligands in organometallic chemistry and homogeneous catalysis. *Chem. Rev.*, 1977, **77**, 313.

W. W. Tomlinson, D. H. Mayo, R. M. Wilson, and J. P. Hooper, The role of ligand steric bulk in new monovalent aluminum compounds. *J. Phys. Chem. A*, 2017, **121**, 4678.

Chapter 6: Circular Dichroism Spectroscopy

Farrukh Vohidov and Andrew R Barron

Introduction

Circular dichroism (CD) spectroscopy is one of few structure assessment methods that can be utilized as an alternative and amplification to many conventional analysis techniques with advantages such as rapid data collection and ease of use. Since most of the efforts and time spent in advancement of chemical sciences are devoted to elucidation and analysis of structure and composition of synthesized molecules or isolated natural products rather than their preparation, one should be aware of all the relevant techniques available and know which instrument can be employed as an alternative to any other technique.

Optical activity

As CD spectroscopy can analyze only optically active species, it is convenient to start the module with a brief introduction of optical activity. In nature almost every life form is handed, meaning that there is certain degree of asymmetry, just like in our hands. One cannot superimpose right hand on the left because they are non-identical mirror images of one another. So are the chiral (handed) molecules, they exist as enantiomers, which mirror images of each other (Figure 6.1). One interesting phenomenon related to chiral molecules is their ability to rotate plane of polarized light. Optical activity property is used to determine specific rotation, α, of pure enantiomer. This feature is used in polarimetry to find the enantiomeric excess, (ee), present in sample.

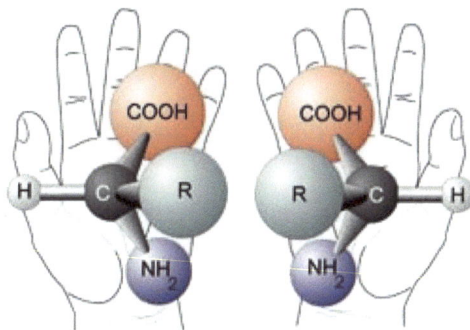

Figure 6.1: Schematic depiction of chirality/handedness of an amino acid.

Circular dichroism

Circular dichroism (CD) spectroscopy is a powerful yet straightforward technique for examining different aspects of optically active organic and inorganic molecules. Circular dichroism has applications in variety of modern research fields ranging from biochemistry to inorganic chemistry. Such widespread use of the technique arises from its essential property of providing structural information that cannot be acquired by other means. One other laudable feature of CD is its being a quick, easy technique that makes analysis a matter of minutes. Nevertheless, just like all methods, CD has a number of limitations, which will be discussed while comparing CD to other analysis techniques.

CD spectroscopy and related techniques were considered as esoteric analysis techniques needed and accessible only to a small clandestine group of professionals. In order to make the reader more familiar with the technique, first of all, the principle of operation of CD and its several types, as well as related techniques will be shown. Afterwards, sample preparation and instrument use will be covered for protein secondary structure study case.

Depending on the light source used for generation of circularly polarized light, there are:
- Far UV CD, used to study secondary structure proteins.
- Near UV CD, used to investigate tertiary structure of proteins.
- Visible CD, used for monitoring metal ion protein interactions.

Principle of operation

In the CD spectrometer the sample is places in a cuvette and a beam of light is passed through the sample. The light (in the present context all electromagnetic waves will be refer to as light) coming from source is subjected to circular polarization, meaning that its plane of polarization is made to rotate either clockwise (right circular polarization) or anti-clockwise (left circular polarization) with time while propagating, see Figure 6.2.

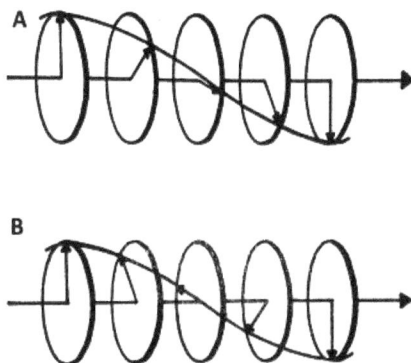

Figure 6.2: Schematic representation of (a) right circularly polarized and (b) left circularly polarized light. Adapted from L. Que, Physical Methods in Bio-inorganic Chemistry Spectroscopy and Magnetism, University Science Books, Sausalito (2000).

The sample is, firstly irradiated with left rotating polarized light, and the absorption is determined by,

$$A = \varepsilon \, c \, l$$

A second irradiation is performed with right polarized light. Now, due to the intrinsic asymmetry of chiral molecules, they will interact with circularly polarized light differently according to the direction of rotation there is going to be a tendency to absorb more for one of rotation directions. The difference between absorption of left and right circularly polarized light is the data, which is obtained from,

$$\Delta A = A_L - A_R = (\varepsilon_L - \varepsilon_R) \, c \, l$$

where ε_L and ε_R are the molar extinction coefficients for left and right circularly polarized light, c is the molar concentration, l is the path length, the cuvette width (in cm). The difference in absorption can be related to difference in extinction, $\Delta\varepsilon$, by,

$$\Delta\varepsilon = \varepsilon_L - \varepsilon_R$$

Usually, due to historical reasons the CD is reported not only as difference in absorption or extinction coefficients but as degree of ellipticity, [θ]. The relationship between [θ] and Δε is given by,

$$[\theta] = 3,298 \, \Delta\varepsilon$$

Since the absorption is monitored in a range of wavelengths, the output is a plot of [θ] versus wavelength or Δε versus wavelength. Figure 6.2 shows the CD spectrum of Δ [Co(en)₃]Cl₃ (Figure 6.3), where en = ethylenediamine (Figure 6.4).

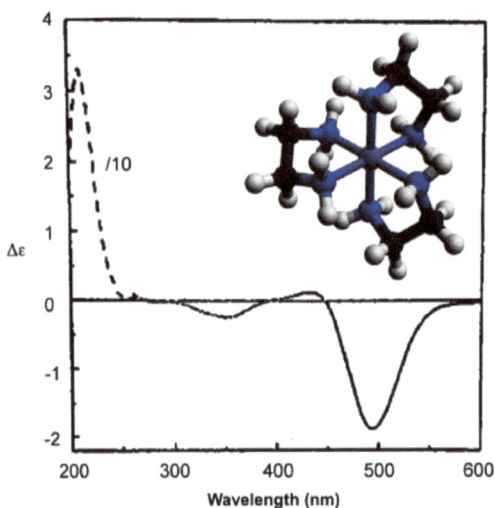

Figure 6.3: CD spectrum of Δ[Co(en)₃]Cl₃ (inset).

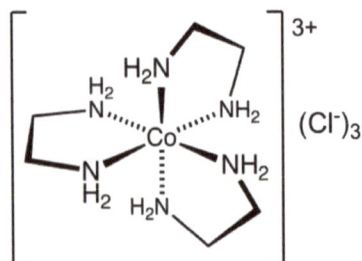

Figure 6.4: Structure of Δ[Co(en)₃]Cl₃, en = ethylenediamine.

Related techniques

Magnetic circular dichroism

Magnetic circular dichroism (MCD) is a sister technique to CD, but there are several distinctions:

- MCD does not require the sample to possess intrinsic asymmetry (i.e., chirality/optical activity), because optical activity is induced by applying magnetic field parallel to light.
- MCD and CD have different selection rules, thus information obtained from these two sister techniques is different. CD is good for assessing environment of the samples' absorbing part while MCD is superior for obtaining detailed information about electronic structure of absorbing part.

MCD is powerful method for studying magnetic properties of materials and has recently been employed for analysis of iron-nitrogen compound, the strongest magnet known. Moreover, MCD and its variation, variable temperature MCD are complementary techniques to Mössbauer spectroscopy and electron paramagnetic resonance (EPR) spectroscopy. Hence, these techniques can give useful amplification to the chapter about Mössbauer and EPR spectroscopy.

Linear dichroism

Linear dichrosim (LD) is also a very closely related technique to CD in which the difference between absorbance of perpendicularly and parallelly polarized light is measured. In this technique the plane of polarization of light does not rotate. LD is used to determine the orientation of absorbing parts in space.

Advantages and limitations of CD

Just like any other instrument CD has its strengths and limits. The comparison between CD and NMR shown in Table 6.1 gives a good sense of capabilities of CD.

CD	NMR
Molecules of any size can be studied	There is size limitation
The experiments are quick to perform; single wave- length measurements require milliseconds.	This is not the case all of the time.
Unique sensitivity to asymmetry in sample's structure.	Special conditions are required to differentiate be- tween enantiomers.
Can work with very small concentrations, by lengthening the cuvette width until discernable absorption is achieved.	There is a limit to sensitivity of instrument.
Timescale is much shorter (UV) thus allowing to study dynamic systems and kinetics.	Timescale is long, use of radio waves gives average of all dynamic systems.
Only qualitative analysis of data is possible.	Quantitative data analysis can be performed to estimate chemical composition.
Does not provide atomic level structure analysis	Very powerful for atomic level analysis, providing essential information about chemical bonds in system.
The observed spectrum is not enough for claiming one and only possible structure	The NMR spectrum is key information for assigning a unique structure.

Table 6.1: A comparison of CD spectroscopy to NMR spectroscopy.

What kind of data is obtained from CD?

One effective way to demonstrate capabilities of CD spectroscopy is to cover the protein secondary structure study case, since CD spectroscopy is well-established technique for elucidation of secondary structure of proteins as well as any other macromolecules. By using CD one can estimate the degree of conformational order (what percent of the sample proteins is in α-helix and/or β-sheet conformation), see Figure 6.5.

Key points for visual estimation of secondary structure by looking at a CD spectrum:
- α-helical proteins have negative bands at 222 nm and 208 nm and a positive band at 193 nm.
- β-helices have negative bands at 218 nm and positive bands at 195 nm.
- Proteins lacking any ordered secondary structure will not have any peaks above 210 nm.

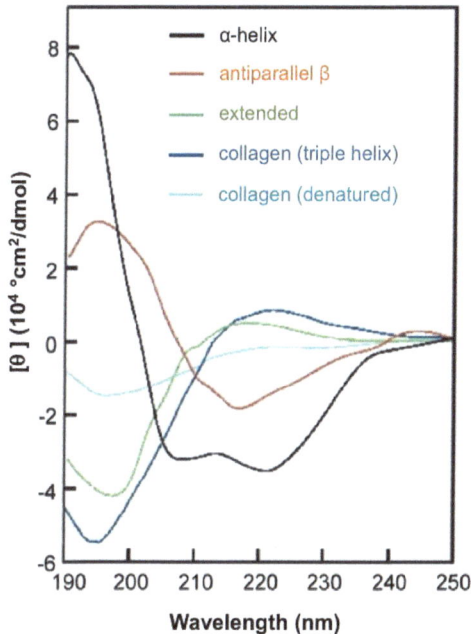

Figure 6.5: CD spectra of samples with representative conformations. Adapted from N. Greenfield, Using circular dichroism spectra to estimate protein secondary structure. *Nat. Protoc.,* **2006, 1, 2876. Copyright: Nature Publishing Group (2006).**

Since the CD spectra of proteins uniquely represent their conformation, CD can be used to monitor structural changes (due to complex formation, folding/unfolding, denaturation because of rise in temperature, denaturants, change in amino acid sequence/mutation, etc.) in dynamic systems and to study kinetics of protein. In other words, CD can be used to perform stability investigations and interaction modeling.

CD instrument

Figure 6.6 shows a typical CD instrument.

Protocol for collecting a CD spectrum

Described is a general procedure for data collection, options (time constant of instrument, wavelength interval, half-bandwidth) can be varied according to needs through the instrument controlling program.

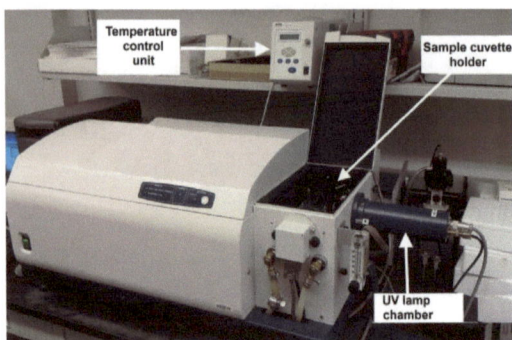

Figure 6.6: A CD instrument.

Sample preparation and starting the instrument

Most of proteins and peptides will require using buffers in order to prevent denaturation. Caution should be shown to avoid using any optically active buffers. Clear solutions are required. CD is taken in high transparency quartz cuvettes to ensure least interference. There are cuvettes available that have path-length ranging from 0.01 cm to 1 cm. Depending on UV activity of buffers used one should choose a cuvette with path-length (distance the beam of light passes through the sample) that compensates for UV absorbance of buffer. Solutions should be prepared according to cuvette that will be used, see Table 6.2.

Cuvette path (cm)	Concentration of sample (mg/mL)
0.01 - 0.02	0.2 1.0
0.1	0.05 0.2
1	0.005 0.01

Table 6.2: Choosing the appropriate cuvette based upon the sample concentration.

Besides, just like salts used to prepare pallets in FT-IR, the buffers in CD will show cutoffs at a certain point in low wavelength region, meaning that buffers start to absorb after certain wavelength. The cutoff values for most of common buffers are known and can be found from manufacturer. Oxygen absorbs light below 200 nm. Therefore, in order to remove interference buffers should be prepared from distilled water or the water should be degassed before use. Another important point is to accurately determine concentration of sample, because concentration should be known for CD data analysis. Concentration of sample can be determined from extinction coefficients, if such are reported in literature also for protein samples quantitative amino acid analysis can be used.

Many CD instrument come bundled with a sample compartment temperature control unit. This is very handy when doing stability and unfolding/denaturation studies of proteins. Check to make sure the heat sink is lled with water. Turn the temperature control unit on and set to chosen temperature.

UV source in CD is very powerful lamp and can generates large amounts of Ozone in its chamber. Ozone significantly reduces the life of the lamp. Therefore, oxygen should be removed before turning on the main lamp (otherwise it will be converted to ozone near lamp). For this purpose, nitrogen gas is constantly flushed into lamp compartment. Let nitrogen flush at least for 15 min. before turning on the lamp.

Collecting spectra for blank, water, buffer background, and sample

- Collect spectrum of air blank (Figure 6.7a). This will be essentially a line lying on×axis of spectrum, zero absorbance.
- Fill the cuvette with water and take a spectrum.
- Water droplets left in cuvette may change concentration of your sample, especially when working with dilute samples. Hence, it is important to thoroughly dry the cuvette. After drying the cuvette, collect spectrum of buffer of exactly same concentration as used for sample (Figure 6.7b). This is the step where buffer is confirmed to be suitable spectrum of the buffer and water should overlap within experimental error, except for low wavelength region where signal-to-noise ratio is low.
- Clean the cuvette as described above and fill with sample solution. Collect the CD spectrum for three times for better accuracy (Figure 6.7c). For proteins multiple scans should overlap and not drift with time.

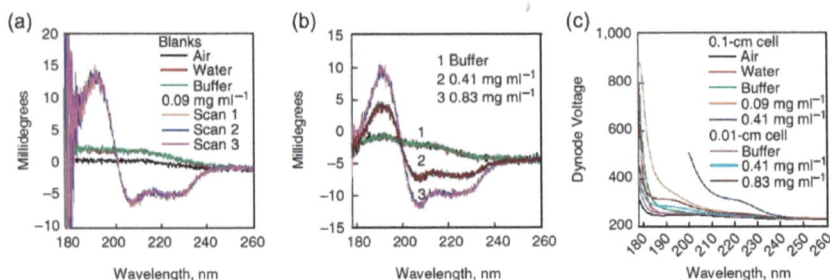

Figure 6.7: CD spectra of blank and water (a), buffer (b), and sample (c). Lysozyme in 10 mM sodium phosphate pH 7. Adapted from N. Greenfield, Using circular dichroism spectra to estimate protein secondary structure. *Nat. Protoc.*, 2006, 1, 2876. Copyright: Nature Publishing Group (2006).

Data handling and analysis

After saving the data for both the spectra of the sample and blank is smoothed using built-in commands of controller software. The smoothed baseline is subtracted from the smoothed spectrum of the sample. The next step is to use software bundles which have algorithms for estimating secondary structure of proteins. Input the data into the software package of choice and process it. The output from algorithms will be the percentage of a particular secondary structure conformation in sample. The data shown in Table 6.3 lists commonly used methods and comparers them for several proteins. The estimated secondary structure is compared to X-ray data, and one can see that it is best to use several methods for best accuracy.

Method		X-ray	LINCOMB	MLR	CONTIN	VARSLC
Lactate de-	α-helix	37	46	63	46	40
hydrogena	β-sheet	14	21	15	7	15
se (Figure 6.7)	Turn	25	15	14	26	17
Chymo-	α-helix	10	15	33	11	24
trypsin	β-sheet	38	25	6	16	0
(Figure 6.8)	Turn	26	10	5	44	42

Table 6.3: Comparison of secondary structure estimation methods. CD methods used 200 nm minimum wavelength. Data from N. Greenfield, Using circular dichroism spectra to estimate protein secondary structure. *Nat. Protoc.*, 2006, 1, 2876.

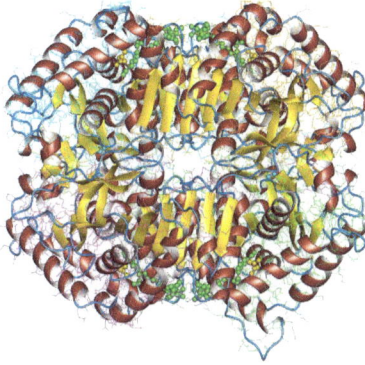

Figure 6.7: Structure of lactate dehydrogenase enzyme.

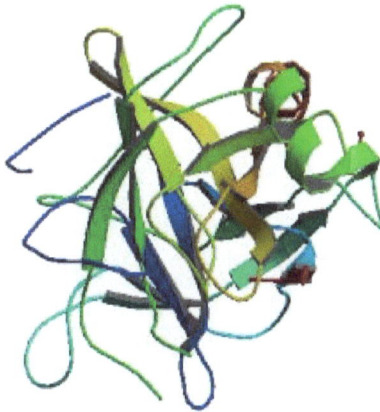

Figure 6.8: Structure of the digestive enzyme chymotrypsin.

Conclusion

What advantages CD has over other analysis methods? CD spectroscopy is an excellent, rapid method for assessing the secondary structure of proteins and performing studies of dynamic systems like folding and binding of proteins. It worth noting that CD does not provide information about the position of those subunits with specific conformation. However, CD outrivals other techniques in rapid assessing of the structure of unknown protein samples and in monitoring structural changes of known proteins caused by ligation and complex formation, temperature change, mutations, denaturants. CD is also widely used to juxtapose fused proteins with wild type counterparts, because CD spectra can tell whether the fused protein retained the structure of wild type or underwent changes.

Bibliography

P. Atkins and J. de Paula, *Elements of Physical Chemistry*, 4th edn., Oxford University Press (2005).

N. Greenfield, Using circular dichroism spectra to estimate protein secondary structure. *Nat. Protoc.*, 2006, **1**, 2876.

N. Greenfield and G.D. Fasman, Computed circular dichroism spectra for the evaluation of protein conformation. *Biochemistry*, 1969, **8**, 4108.

G. Holzwarth and P. Doty, The ultraviolet circular dichroism of polypeptides. *J. Am. Chem. Soc.*, 1965, **87**, 218.

Z. Lin and X. M. Pan, Accurate prediction of protein secondary structural content. *J. Protein Chem.*, 2001, **20**, 217.

A. Perczel, K. Park, and G. D. Fasman, Analysis of the circular dichroism spectrum of proteins using the convex constraint algorithm: A practical guide. *Anal Biochem.*, 1992, **203**, 83.

L. Que, Physical *Methods in Bioinorganic Chemistry Spectroscopy and Magnetism*, University Science Books, Sausalito (2000).

M. Rutherfurd and M. Dunn, Quantitative amino acid analysis. *Curr. Protoc. Protein Sci.*, 2011, **63**, 3.2.1.

N. Sreerama and R. W. Woody, Estimation of protein secondary structure from circular dichroism spectra: comparison of CONTIN, SELCON, and CDSSTR methods with an expanded reference set. *Anal. Biochem.*, 2000, **287**, 252.

Chapter 7: Analysis of Liquid Crystal Phases using Polarized Optical Microscopy

Changsheng Xiang and Andrew R. Barron

Introduction

The properties of liquid crystals (LCs) were first discovered by botanical physiologist Friedrich Reinitzer (Figure 7.1), who examined the physico-chemical properties of various derivatives of cholesterol. Reinitzer perceived that color changes in a derivative cholesteryl benzoate (Figure 7.2) observed by others were not the most peculiar feature. He found that cholesteryl benzoate does not melt in the same manner as other compounds but has two melting points. At 145.5 °C it melts into a cloudy liquid, and at 178.5 °C it melts again and becomes clear. He also observed that the phenomenon is reversible. Reinitzer wrote to Otto Lehmann (Figure 7.3), who had a rare microscope in his laboratory that allowed the observation of crystallites in the cloudy liquid above 145.5 °C.

Figure 7.1: Austrian botanist and chemist Friedrich Richard Reinitzer (1857 - 1927).

Reinitzer had discovered and described three important features of cholesteric liquid crystals (the name coined by Lehmann): the existence of two melting points, the reflection of circularly polarized light, and the ability to rotate the polarization direction of light. However, he did not peruse his studies further, but Lehmann started a systematic study that was later continued by Daniel Vorländer who went on to synthesize most of the LCs known until 1935.

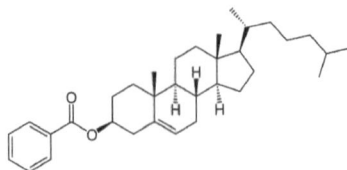

Figure 7.2. Chemical structure of cholesteryl benzoate molecule

Figure 7.3: German physicist Otto Lehmann (1855 - 1922).

Figure 7.4: German chemist Daniel Vorländer (1867 - 1941).

Liquid crystal phases

Liquid crystals are a state of matter that has the properties between solid crystal and common liquid. There are basically three different types of liquid crystal phases:

- Thermotropic liquid crystal phases are dependent on temperature.
- Lyotropic liquid crystal phases are dependent on temperature and the concentration of LCs in the solvent.
- Metallotropic LCs are composed of organic and inorganic molecules, and the phase transition not only depend on temperature and concentration, but also depend on the ratio between organic and inorganic molecules.

Thermotropic LCs are the most widely used one, which can be divided into five categories:

- Nematic phase in which rod-shaped molecules have no positional order, but they self-align to have long-range directional order with their long axes roughly parallel (Figure 7.5a).
- Smactic phase where the molecules are positionally ordered along one direction in well-defined layers oriented either along the layer normal (smectic A) or tilted away from the layer normal (smectic C), see Figure 7.5b.
- Chiral phase which exhibits a twisting of the molecules perpendicular to the director, with the molecular axis parallel to the director (Figure 7.5c).

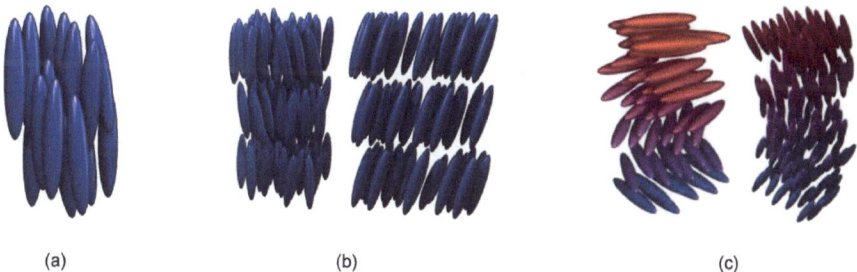

(a)   (b)   (c)

Figure 7.5: Schematic representations of (a) a nematic LC phase, (b) smactic LC phases oriented along (left) and away (right) from the normal of the layer, and (c) a chiral LC phase.

- Blue phase having a regular three-dimensional cubic structure of defects with lattice periods of several hundred nanometers, and thus they exhibit selective Bragg reflections in the wavelength range of light (Figure 7.6).
- Discotic phase in which disk-shaped LC molecules can orient themselves in a layer-like fashion (Figure 7.7).

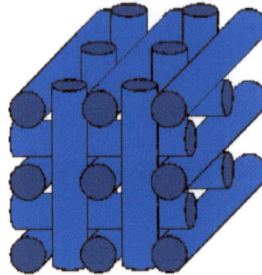

Figure 7.6: A schematic representation of the ordered structure of a blue LC phase.

(a)   (b)

Figure 7.7: Schematic representations of (a) a discotic nematic LC phase and (b) a discotic columnar LC phase.

Thermotropic LCs are very sensitive to temperature. If the temperature is too high, thermal motion will destroy the ordering of LCs, and push it into a liquid phase. If the temperature is too low, thermal motion is hard to perform, so the material will become crystal phase.

The existence of liquid crystal phase can be detected by using polarized optical microscopy, since liquid crystal phase exhibits its unique texture under microscopy. The contrasting areas in the texture correspond to domains where LCs are oriented towards different directions.

Polarized optical microscopy

Polarized optical microscopy is typically used to detect the existence of liquid crystal phases in a solution. A polarizer is a filter that only permits the light oriented in a specific direction with its polarizing direction to pass through. There are two polarizers in a polarizing optical microscope (POM) (Figure 7.8) and they are designed to be oriented at right angle to each other, which is termed as cross polar.

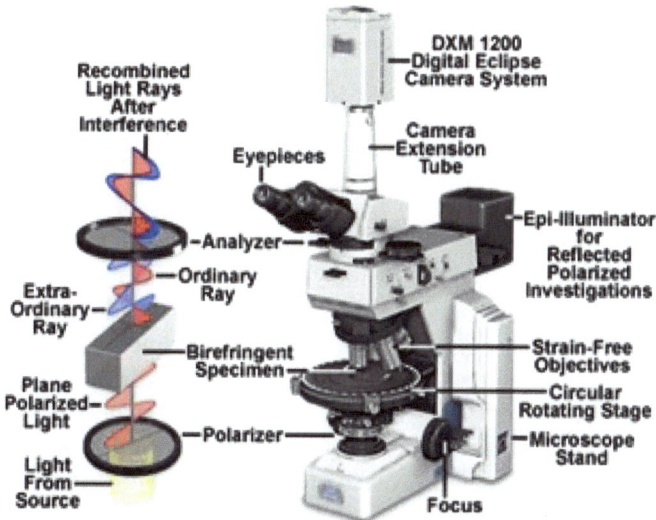

Figure 7.8: The basic configuration of polarized optical microscope. Copyright: Nikon Corporation.

The fundamental of cross polar is illustrated in Figure 7.9, the polarizing direction of the first polarizer is oriented vertically to the incident beam, so only the waves with vertical direction can pass through it. The passed wave is subsequently blocked by the second polarizer, since this polarizer is oriented horizontally to the incident wave.

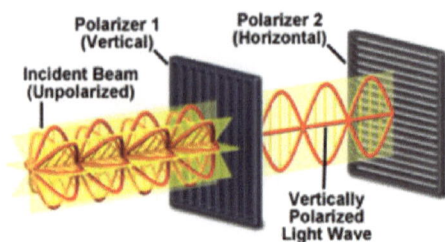

Figure 7.9: A schematic representation of the polarization of light waves. Copyright: Nikon Corporation.

Theory of birefringence

A birefringent or doubly-refracting, sample has a unique property that it can produce two individual wave components while one wave passes through it, those two components are termed as ordinary and extraordinary waves. Figure 7.10 is an illustration of a typical construction of Nicol polarizing prism, as we can see, the non-polarized white light is split into two rays as it passes through the prism. The one travels out of the prism is called ordinary ray, and the other one is called extraordinary ray. So, if we have a birefringent specimen located between the polarizer and analyzer, the initial light will be separated into two waves when it passes though the specimen. After exiting the specimen, the light components become out of phase, but are recombined with constructive and destructive interference when they pass through the analyzer. Now the combined wave will have elliptically or circularly polarized light wave, see Figure 7.11, image contrast arises from the interaction of plane-polarized light with a birefringent specimen so some amount of wave will pass through the analyzer and give a bright domain on the specimen.

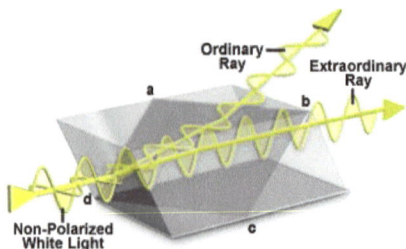

Figure 7.10: A schematic representation of a Nicol polarzing prism. Copyright: Nikon Corporation.

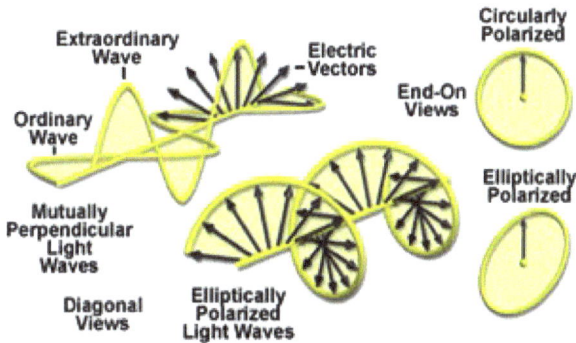

Figure 7.11: A schematic representation of elliptically and circularly polarized light waves. Copyright: Nikon Corporation.

Liquid crystal display

The most common application of LCs is in liquid crystals displays (LCD). Figure 7.12 is a simple demonstration of how LCD works in digit calculators. There are two crossed polarizers in this system, and liquid crystal (cholesteric spiral pattern) sandwich with positive and negative charging is located between these two polarizers. When the liquid crystal is charged, waves can pass through without changing orientations. When the liquid crystal is out of charge, waves will be rotated 90° as it passes through LCs so it can pass through the second polarizer. There are seven separately charged electrodes in the LCD, so the LCD can exhibit different numbers from 0 to 9 by adjusting the electrodes. For example, when the upper right and lower left electrodes are charged, we can get 2 on the display.

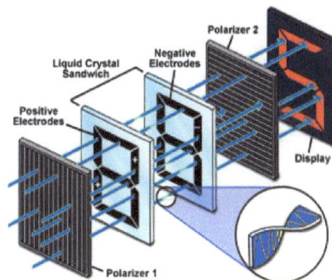

Figure 7.12: Demonstration of a seven-segment liquid crystal display. Copyright: Nikon Corporation.

Microscope images of liquid crystal phase

The first order retardation plate

The first order retardation plate is frequently utilized to determine the optical sign of a birefringent specimen in polarized light microscopy. The optical sign includes positive and negative. If the ordinary wavefront is faster than the extraordinary wavefront, the specimen displays positive birefringence. Conversely a negative birefringence will be detected if the ordinary wavefront is slower than the extraordinary wavefront. In addition, the retardation plate is also useful for enhancing contrast in weakly birefringent specimens. Figure 7.13 shows the effect of first order retardation plate on the contrast of birefringence. The birefringence is so weak that the morphology on the edge of the tissue is hard to image, Figure 7.13a. When a first order retardation plate is added, the structure of the cell become all apparent compared with the one without retardation plate, Figure 7.13b.

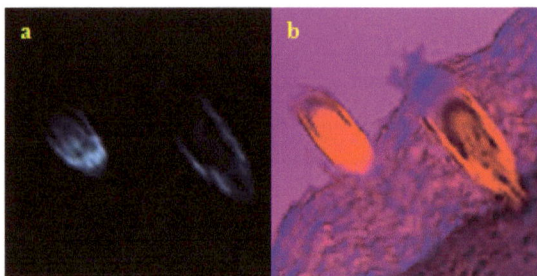

Figure 7.13: Microscope images of thin section of human tongue (a) without first order retardation plate and (b) with first order retardation plate. Copyright: Olympus.

Images of liquid crystal phases

Figure 7.14 shows the images of liquid crystal phases from different specimens. First order retardation plates are utilized in all of these images. Apparent contrasts are detected here in the image which corresponds to the existence of liquid crystal phase within the specimen.

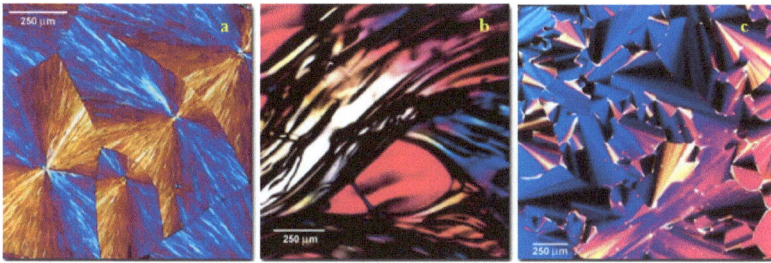

Figure 7.14: Microscope images in polarized light with a first-order retardation plate inserted between the specimen and analyzer: (a) polyethylene glycol, (b) polycarbonate, and (c) liquid crystalline DNA. Copyright from Nikon.

The effect of rotation of the polarizer

The effect of the angle between horizontal direction and polarizer transmission axis on the appearance of liquid crystal phase may be analyzed. In Figure 7.15 is shown images of an ascorbic acid (Figure 7.16) sample under cross polar mode. When the polarizer rotates from 0° to 90°, big variations on the shape of bright domains and domain colors appear due to the change of wave vibrating directions. By rotating the polarizer, we can have a comprehensive understanding of the overall texture.

Figure 7.15: Cross polarized Microscope images of ascorbic acid specimen with polarizer rotation of (a) 0°, (b) 45°, and (c) 90°. Copyright: Nikon Corporation.

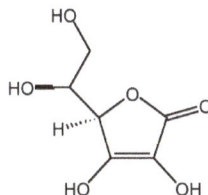

Figure 7.16: The structure of ascorbic acid.

Bibliography

R. Weaver, Rediscovering polarized light microscopy. *Am. Lab.*, 2003, **35**, 55.

F. Massoumian, R. Juskaitis, M. A. Neil, and T. Wilson, Quantitative polarized light microscopy. *J. Microsc.*, 2003, **209**, 13.

R. Oldenbourg, New views on polarization microscopy. *Nature*, 1996, **381**, 811.

Chapter 8: Protein Analysis using Electrospray Ionization Mass Spectroscopy

Wilhelm Kienast and Andrew R. Barron

Introduction

Electrospray ionization-mass spectrometry (ESI-MS) is an analytical method that focuses on macromolecular structural determination. The unique component of ESI-MS is the electrospray ionization. The development of electrospraying, the process of charging a liquid into a fine aerosol, was completed in the 1960's when Malcolm Dole (Figure 8.1) demonstrated the ability of chemical species to be separated through electrospray techniques.

Figure 8.1: American chemist Malcolm Dole (on right) (1903 - 1990).

With this important turn of events, the combination of ESI and MS was feasible and was later developed by John B. Fenn (Figure 8.2), as a functional analytical method that could provide beneficial information about the structure and size of a protein. Fenn shared the Nobel Prize in 2002, with Koichi Tanaka (Figure 8.3) and Kurt Wüthrich (Figure 8.4) for the development of ESI-MS.

ESI-MS is the process through which proteins, or macromolecules, in the liquid phase are charged and fragmented into smaller aerosol droplets. These aerosol droplets lose their solvent and propel the charged fragments into the gas phase in several components that vary by charge. These components can then be detected by a mass spectrometer. The recent boom and development of ESI-MS is attributed to its benefits in characterizing and analyzing macromolecules, specifically biologically important macromolecules such as proteins.

Figure 8.2: American chemist John Bennett Fenn (1917 - 2010) shared the Nobel Prize for his work in ESI-MS and other identification and structural analyses of biological molecules.

Figure 8.3: Japanese chemist and Nobel laureate Koichi Tanaka (1959 -).

Figure 8.4: Swiss chemist and Nobel laureate Kurt Wüthrich (1938 -).

How does ESI-MS function?

ESI-MS is a process that requires the sample to be in liquid solution, so that tiny droplets may be ionized and analyzed individually by a mass spectrometer. The following delineates the processes that occur as relevant to Figure 8.5.

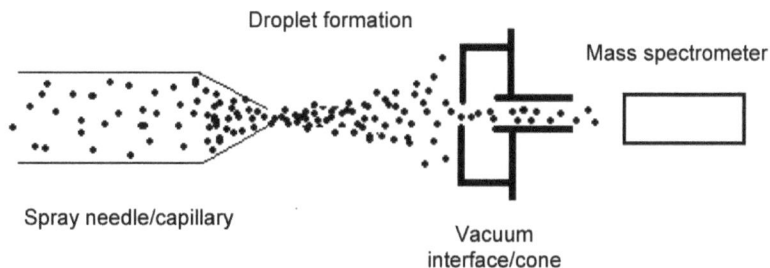

Figure 8.5: The process of ESI-MS. A focus on the capillary spray needle and the generation of aerosol particles.

Spray needle/capillary

The liquid solution of the desired macromolecule is introduced into the system through this needle. The needle is highly charged via an outside voltage source that maintains the charge constant across the needle. The normal charge for a needle is approximately 2.5 to 4 kV. The voltage causes the large droplets to fragment into small droplets based on charge that is accumulated from the protein constituent parts, and the liquid is now in the gas phase.

Droplet formation

The droplets that are expelled from the needle are smaller than initially, and as a result the solvent will evaporate. The smaller droplets then start increasing their charge density on the surface as the volume decreases. As the droplets near the Rayleigh limit, Coulombic interactions of the droplet equal the surface tension of the droplet, a Coulombic explosion occurs that further breaks the droplet into minute fractions, including the isolated analyte with charge.

Vacuum interface/cone

This portion of the device allows for the droplets to align in a small trail and pass through to the mass spectrometer. Alignment occurs because of the similarity and differences in charges amongst all the droplets. All the droplets are ionized to positive charges through addition of protons to varying basic sites on the droplets, yet all the charges vary in magnitude dependent upon the number of basic sites available for protonation. The receiving end or the cone has the opposite charge of the spray needle, causing an attraction between the cone and the droplets.

Mass spectrometer

The charged particles then reach the mass spectrometer and are deflected based on the charge of each particle. Deflection occurs by the quadrupole magnet of the mass spectrometer. The different deflection paths of the ions occur due to the strength of the interaction with the magnetic field. This leads to various paths based on a mass/charge (m/z) ratio. The particles are then read by the ion detector, as they arrive, providing a spectrum based on m/z ratio.

What data is provided by ESI-MS?

As implied by the name, the data produced from this technique is a mass spectrometry spectrum. Without delving too deeply into the topic of mass spectrometry, which is out of the true scope of this module, a slight explanation will be provided here. The mass spectrometer separates particles based on a magnetic field created by a quadrupole magnet. The strength of the interaction varies on the charge the particles carry. The amount of deflection or strength of interaction is determined by the ion detector and quantified into a mass/charge (m/z) ratio. Because of this information, determination of chemical composition or peptide structure can easily be managed as is explained in greater detail in the following section.

Interpretation of a typical MS spectrum

Interpreting the mass spectrometry data involves understanding the m/z ratio. The knowledge necessary to understanding the interpretation of the spectrum is that the peaks correspond to portions of the whole molecule. That is to say, hypothetically, if you put a human body in the mass spectrometer, one peak would coincide with one arm, another peak would coincide with the arm and the abdomen, etc. The general idea behind these peaks, is that an overlay would paint the entire picture, or in the case of the hypothetical example, provide the image of the human body. The m/z ratio defines these portions based on the charges carried by them. Thus, the terminology of the mass/charge ratio. The more charges a portion of the macromolecule or protein holds, the smaller the m/z ratio will be and the farther left it will appear on the spectrum. The fundamental concept behind interpretation involves understanding that the peaks are interrelated, and thus the math calculations may be carried out to provide relevant information of the protein or macromolecule being analyzed.

Calculation of m/z of the MS spectrum peaks

As mentioned above, the pertinent information to be obtained from the ESI-MS data is extrapolated from the understanding that the peaks are interrelated. The steps for calculating the data are as follow:

- Determine which two neighboring peaks will be analyzed.
- Establish the first peak (the one farthest left) as the peak with the greatest m/z ratio. This is mathematically defined as our $z+1$ peak.
- Establish the adjacent peak to the right of our first peak as the peak with the lower m/z ratio. This is mathematically our z peak.
- Our $z+1$ peak will also be our $m+1$ peak as the difference between the two peaks is the charge of one proton. Consequently, our z peak will be defined as our m peak.
- Solve both equations for m to allow for substitution. Both sides of the equation should be in terms of z and can be solved.
- Determine the charge of the z peak and subsequently, the charge of the $z+1$ peak.
- Subtract one from the m/z ratio and multiply the m/z ratio of each peak by the previous charges determined to obtain the mass of the protein or macromolecule.
- Average the results to determine the average mass of the macromolecule or protein.

Example

Determine which two neighboring peaks will be analyzed from the MS (Figure 8.6) as the $m/z = 5$ and $m/z = 10$ peaks.

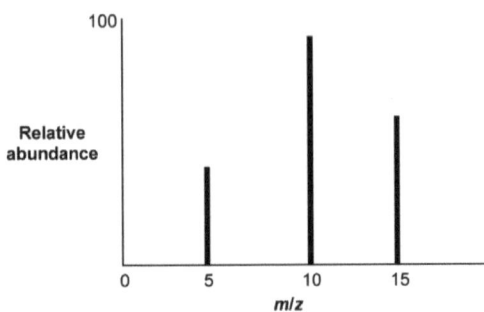

Figure 8.6: Hypothetical mass spectrometry data; not based on of any particular compound. The example steps are based on of this spectrum.

- Establish the first peak (the one farthest left in Figure 8.6) as the $z + 1$ peak (i.e., $z + 1 = 5$).
- Establish the adjacent peak to the right of the first peak as the z peak (i.e., $z = 10$).
- Establish the peak ratios,

$$\frac{m+1}{z+1} = 5$$

$$\frac{m}{z} = 10$$

- Solve the ratios for m:

$$m = 5z + 4$$

$$m = 10z$$

- Substitute one equation for m:

$$5z + 4 = 10z$$

- Solve for z:

$z = 4/5$

- Find $z + 1$:

$z + 1 = 9/5$

- Find average molecular mass by subtracting the mass by 1 and multiplying by the charge,

$(10 - 1)(4/5) = 7.2$

$(5 - 1)(9/5) = 7.2$

- Hence, the average mass = 7.2.

Sample preparation

Samples for ESI-MS must be in a liquid state. This requirement provides the necessary medium to easily charge the macromolecules or proteins into a fine aerosol state that can be easily fragmented to provide the desired outcomes. The bene t to this technique is that solid proteins that were once difficult to analyze, like metallothionein, can dissolved in an appropriate solvent that will allow analysis through ESI-MS. Because the sample is being delivered into the system as a liquid, the capillary can easily charge the solution to begin fragmentation of the protein into smaller fractions Maximum charge of the capillary is approximately 4 kV. However, this amount of charge is not necessary for every macromolecule. The appropriate charge is dependent on the size and characteristic of the solvent and each individual macromolecule. This has allowed for the removal of the molecular weight limit that was once held true for simple mass spectrometry analysis of proteins. Large proteins and macromolecules can now easily be detected and analyzed through ESI-MS due to the facility with which the molecules can fragment.

Related techniques

A related technique that was developed at approximately the same time as ESI-MS is matrix assisted laser desorption/ionization mass spectrometry

(MALDI-MS). This technique that was developed in the late 1980's as wells, serves the same fundamental purpose; allowing analysis of large macromolecules via mass spectrometry through an alternative route of generating the necessary gas phase for analysis. In MALDI-MS, a matrix, usually comprised of crystallized 3,5-dimethoxy-4-hydroxycinnamic acid (Figure 8.7), water, and an organic solvent, is used to mix the analyte, and a laser is used to charge the matrix. The matrix then co-crystallizes the analyte and pulses of the laser are then used to cause desorption of the matrix and some of the analyte crystals with it, leading to ionization of the crystals and the phase change into the gaseous state. The analytes are then read by the tandem mass spectrometer. Table 8.1 directly compares some attributes between ESI-MS and MALDI-MS. It should be noted that there are several variations of both ESI-MS and MALDI-MS, with the methods of data collection varying and the piggy-backing of several other methods (liquid chromatography, capillary electrophoresis, inductively coupled plasma mass spectrometry, etc.), yet all of them have the same fundamental principles as these basic two methods.

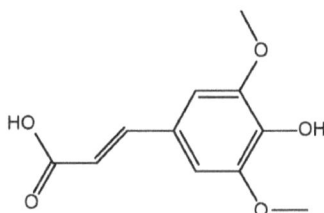

Figure 8.7: Structure of 3,5-dimethoxy-4-hydroxycinnamic acid.

Experimental details	ESI-MS	MALDI-MS
Starting analyte state	Liquid	Liquid/solid
Method of ionization	Charged capillary needle	Matrix laser desorption
Final analyte state	Gas	Gas
Quantity of protein needed	1 μL	1 μL
Spectrum method	Mass spectrometry	Mass spectrometry

Table 8.1: Comparison of the general experimental details of ESI-MS and MALDI-MS.

Problems with ESI-MS

ESI-MS has proven to be useful in determination of tertiary structure and molecular weight calculations of large macromolecules. However, there are still several problems incorporated with the technique and macromolecule analysis. One problem is the isolation of the desired protein for analysis. If the protein is unable to be extracted from the cell, this is usually done through gel electrophoresis, there is a limiting factor in what proteins can be analyzed. Cytochrome c (Figure 8.8) is an example of a protein that can be isolated and analyzed but provides an interesting limitation on how the analytical technique does not function for a completely effective protein analysis. The problem with cytochrome c is that even if the protein is in its native confirmation, it can still show different charge distribution. This occurs due to the availability of basic sites for protonation that are consistently exposed to the solvent. Any slight change to the native conformation may cause basic sites, such as in cytochrome c to be blocked causing different m/z ratios to be seen. Another interesting limitation is seen when inorganic elements, such as in metallothioneins proteins that contain zinc, are analyzed using ESI-MS. Metallothioneins have several isoforms that show no consistent trend in ESI-MS data between the varied isoforms. The marked differences occur due to the metallation of each isoform being different, which causes the electrospraying and as a result protonation of the protein to be different. Thus, incorporation of metal atoms in proteins can have various effects on ESI-MS data due to the unexpected interactions between the metal center and the protein itself.

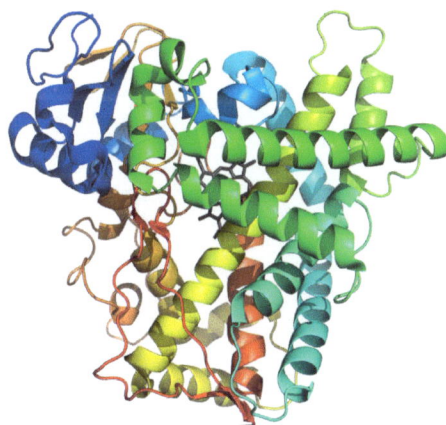

Figure 8.8: The 3-D structure of human cytochrome P450 2A13, a sub class of human cytochrome c.

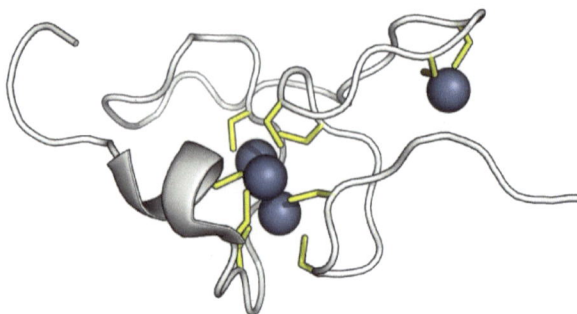

Figure 8.9: The structure of the beta-E-domain of wheat Ec-1 metal-lothionein (PDB:2KAK), with cysteines in yellow and zinc ions in purple.

Bibliography

L. Konermann and D. J. Douglas, Acid-induced unfolding of cytochrome *c* at different methanol concentrations: electrospray ionization mass spectrometry specifically monitors changes in the tertiary structure. *Biochemistry*, 1997, **36**, 12296.

S. Pérez-Rafael, S. Atrian, M. Capdevila, and Ò. Palacios, Differential ESI-MS behaviour of highly similar metallothioneins. *Talanta*, 2011, **83**, 1057.

D. L. Nelson and M. M. Cox, *Lehninger Principles of Biochemistry*, 5th edn., W. H. Freeman and Company, New York (2008).

A. Prange and D. Profrock, Application of CE–ICP–MS and CE–ESI–MS in metalloproteomics: challenges, developments, and limitations. *Anal. Bioanal. Chem.*, 2005, **383**, 372.

www.ingramcontent.com/pod-product-compliance
Lightning Source LLC
Chambersburg PA
CBHW041932220326
41598CB00055BA/28